Heterogeneous Catalytic Science

Author

R. D. Srivastava
University of Delaware
Center for Catalytic Science and Technology
Department of Chemical Engineering
Newark, Delaware

CRC Press, Inc.
Boca Raton, Florida

Library of Congress Cataloging-in-Publication Data

Srivastava, R. D.
Heterogeneous catalytic science/author, R. D. Srivastava.
p. cm.
Bibliography: p.
Includes index.
ISBN 0-8493-6430-2
1. Heterogeneous catalysis. I. Title.
QD505.S697 1988
541.395--dc19

87-21924
CIP

This book represents information obtained from authentic and highly regarded sources. Reprinted material is quoted with permission, and sources are indicated. A wide variety of references are listed. Every reasonable effort has been made to give reliable data and information, but the author and the publisher cannot assume responsibility for the validity of all materials or for the consequences of their use.

Direct all inquiries to CRC Press, Inc., 2000 Corporate Blvd., N.W., Boca Raton, Florida, 33431.

International Standard Book Number 0-8493-6430-2

Library of Congress Card Number 87-21924
Printed in the United States

PREFACE

Catalysis encircles a variety of fields, and its applications are indeed complex. The rapid progress of the many branches of chemistry and chemical engineering which contribute to catalysis tend to widen the communication gap between the scientists and the engineers involved in catalysis, and it is the author's intention to narrow this gap by integrating the areas of surface science, applied catalysis, and chemical reaction engineering. Emphasizing an interdisciplinary viewpoint, this book is designed to stimulate new progressive ideas throughout this broad science, offering a representative cross-section of surface and applied catalyses and some insight into the catalytic practice.

The opening chapter deals with the concepts of heterogeneous catalytic science and chemical reaction engineering. An analysis of several key techniques useful for characterizing heterogeneous catalysts is then provided. This chapter is intended to serve as a guide to methods of catalyst surface characterization and to provide a description of the special advantages and disadvantages, and success of specific methods.

The remaining four chapters describe several case histories of important catalytic processes which serve to illustrate the rapidly accelerating development of catalysis science. The engineering subjects are introduced as they arise in this context. Each chapter is concerned with an industrial process or class of processes, namely, selective oxidation of hydrocarbons, epoxidation of ethene, hydrogenation of carbon monoxide, and catalytic reforming. It is hoped that these reviews will demonstrate the complexity of industrial catalysts, and show a need of further and more intimate interaction between academic and industrial scientists.

This book is so structured that it will complement the existing texts on chemical reaction engineering. The present document is also intended to be particularly useful to scientists and engineers who are engaged in research or development, or who guide the industrial processes in a given specific area of catalysis or surface science. It is the author's intent to provide a significant overview of research and development trends in the areas of large commercial importance. By combining fundamental work and process research with the current state of commercial technology trends, the future directions of the catalytic industry can be deduced.

The book then is not meant to be comprehensive, but a guide to fairly rugged territory.

The manuscript was prepared while I was a Visiting Professor in the Department of Chemical Engineering at the University of Delaware. This evolved partly from notes for a graduate course taught at the Indian Institute of Technology, Kanpur.

Without the help and criticisms of my colleagues, and the comments of industrial chemists and engineers, this book could not have been written. Much of what has been covered about the generality of epoxidation catalysis and carbon monoxide hydrogenation was influenced by the ideas of my colleagues M. A. Barteau and D. L. Cresswell of ICI, U.K.; J. D. Burrington of Standard Oil (Chapter 3), and M. J. Kelley of DuPont (Chapter 6) all of whom served admirably in reading and commenting on the chapter as indicated. T. R. Hughes, formerly of Chevron, offered invaluable comments on the catalytic reforming.

I am grateful to the former department chairman, S. I. Sandler, for his encouragement and stimulation during the preparation of the manuscript. Special thanks go to A. Beris, who assisted me in several ways in the early planning stages of this book. I am also indebted to the many other colleagues, especially to the department chairman, T.W.F. Russell, and to the director of the center, B. C. Gates, who have assisted in general ways.

R. D. Srivastava

THE AUTHOR

Rameshwar D. Srivastava, Ph.D., is currently affiliated with the Burns and Roe Corporation located at the Pittsburgh Energy Technology Center, Bruceton, Pennsylvania. Before this, he was a Visiting Professor in the Chemical Engineering Department at the University of Delaware for 2 years and a Professor of Chemical Engineering at the Indian Institute of Technology, Kanpur, for 13 years. There he assumed the position of Department Chairman. While at IIT, he was also a consultant for Space Sciences, Inc., Monrovia, California over a 12-year period. Of further note, Dr. Srivastava has been a Guest Professor at the Swiss Federal Institute of Technology, Zurich.

Dr. Srivastava received his Ph.D. (1967) in chemical engineering from the Technical University of Nova Scotia, and did postdoctoral research at the University of Toronto. Dr. Srivastava was recepient of the IIChEs' Herdillia Award for Excellence in Basic Research in Chemical Engineering. He has published over 90 research papers relating to catalysis and reaction engineering, and high temperature materials.

DEDICATION

To A. B. Metzner for inspiring and encouraging

TABLE OF CONTENTS

Chapter 1

CONCEPTS OF CATALYTIC SCIENCE AND KINETIC ANALYSES

I. INTRODUCTION

Catalysis is central to the global problem related to energy, resources, and environment. Catalysis is the heart of the chemical and petroleum industry. The success of the chemical industry is based largely on catalyst technology. The development of new and improved catalysts and catalytic processes is vitally important to the industry.

The field of catalytic science is multidimensional and multidisciplinary. Modern access to research and development in this area requires an integrated approach based on different established practices of science and engineering. Since the working of a practical system is so complex, a reliable and reproducible study on a molecular level must be carried out through the use of model systems. The model system so chosen should be comparable to the real one.

Industrial catalysis is mainly influenced by science. Catalysis may be realized both homogeneously and heterogeneously, the most common industrial processes being heterogeneous catalysis by a solid in contact with the reactant — gas.

II. CONCEPTS

According to the transition-state theory, chemical reaction velocity is determined by the free energy of formation of transition complex postulated to exist between reactants and products. Catalysis is that process in which the catalytic agent — 'catalyst' — aids the attainment of chemical equilibrium by reducing the free energy of the transition-complex formation in the reaction path.

Many chemical reactions lead to the formation of several different, but all thermodynamically feasible, products. A selective catalyst may alter the individual steps to different degrees, with the consequence that overall reaction yield or selectivity is affected. Thus, the blocking of reaction pathways which lead to the formation of undesired molecules is as important an attribute of a good catalyst as the lowering of the potential energy barrier along the reaction pathway to the desired product molecule.

Activity is reported in various ways, generally as conversion for specified conditions and volume of catalyst or as relative percentage vs. a standard catalyst. During the last 2 decades, for research or fundamental use, turnover number, which describes the number of molecules produced per surface metal atom per second, has been widely accepted as an appropriate unit to express the activity of a metal for catalyzing a reaction. Not only does such a unit of catalytic activity make it more meaningful to compare various metals for the same reaction conditions, but it also makes it useful in comparing the catalytic activity as a function of metal crystallite size, support, or method of catalyst preparation. Chemisorption measurements are used to calculate turnover numbers. Similarly, the velocity of a catalyzed reaction is described in terms of a 'turnover rate', which reveals how many product molecules are formed per second at a given temperature and pressure.

A surface, in a strict sense, is an infinitely thin gas-solid interface. In catalytic science, the topmost layer of a solid catalyst is referred to as the surface. Since such a layer has a thickness of at least 1 atomic diameter, it is also called a thin film or thin (mono) layer.[1,2] The catalytic action occurs at specific sites on solid surfaces, frequently called 'active sites'. An 'active site' is a site consisting of a coordinative unsaturated atom on which reactants or intermediates can be adsorbed during the sequence of reaction steps. Similarly, the expression 'active center' is used for a group of surface atoms that together bind one adsorbant entity.

A 'reactive ensemble' is a group of active sites involved in a particular reaction step; a 'reactive aggregate' is the system of reactive ensembles which is required for the formation of one product molecule.

Since catalytic activity may be proportional to the concentration of active sites, supports of high area are commonly employed. Bulk metals are used in a few processes; for example, iron is used as a catalyst in the ammonia synthesis process, and Pt-Rh gauge is used in ammonia oxidation for nitric acid manufacture. But by far, the largest applications are of supported metals.

Dispersion on a support provides for the most efficient utilization of metal, since small crystallites (1.0 to 10 nm) expose a large fraction of their atoms on the surface where they are available to reactants. Metal crystallites are usually supported on inorganic oxides of high surface area to provide stability and prevent sintering and metal surface-area loss. Applications of supported metals (with typical catalysts) include catalytic reforming of petroleum hydrocarbons (Pt/Al_2O_3, Pt-Re/Al_2O_3, and Pt-Ir/Al_2O_3), hydrocracking of petroleum hydrocarbons (Pt/zeolite and Pd/zeolite), hydrogenation of fats, hydrocarbons, sugars, etc. (Pd/Al_2O_3, Ni/kieselguhr, and Ni/Al_2O_3), and isomerization of hydrocarbons (Pt/zeolite, Pd/zeolite, and Pt/chlorinated alumina). The largest volume user of supported metal catalysts is the automotive industry in the catalytic muffler utilizing Pt-Pd/Al_2O_3 to oxidize hydrocarbons and carbon monoxide.

The catalysts are usually prepared by finely dispersing the active metal or compound on the supports. The support may have the acid site characteristics in the catalytic process for it may endow the dispersed catalyst with certain properties (state of oxidation and valence) and may exhibit to adsorb reactants and/or atomic species dissociated by the deposited catalytic agents.[3] For example, in the supported Pt-catalyzed reforming, the hydrogenation and dehydrogenation activities are associated with Pt component, while the activity for acid catalysis is associated with the presence of acidic sites on the surface of the support — Al_2O_3.

Total area of the catalyst formulation is routinely secured by physical adsorption data in terms of the Brunauer-Emmett-Teller (BET area) equation. It takes multilayer adsorption into account and is a standardized one which has been reproduced by different workers in different laboratories with considerable accuracy and reproducibility.

The catalytic activity and selectivity of supported metal catalysts depend on the percentage dispersion of the metallic phase, and a knowledge of intrinsic value of catalyst surface-atom concentration is of single importance. The catalyst dispersion is the number of surface metal atoms per total number of metal atoms. Only those sites exposed to reactants as dispersion value of unity (all of the atoms are on surface) are generally considered the catalytic ideal, while a value of near zero (typical of bulk specimen, such as Pt wire) is generally a catalytic undesirable. Chemisorption of gases like hydrogen, oxygen, and carbon monoxide is the most common technique for the measurement of metal dispersion. For example, the stoichiometrics of chemisorption of oxygen and hydrogen and of the H_2-O_2, or O_2-H_2 titrations have been confirmed and accepted for Pt-Al_2O_3 catalysts.[4]

If more than one metal is dispersed on the carrier, the catalysts have much more complex gas adsorption behavior than when only one component is present, because the magnitude of gas uptake may well be a function of both the composition of the metallic surface with respect to the two components and of the total metallic surface area. For such bimetallic catalysts, the state of metal dispersion can also reveal whether they exist as alloys or separate and distinct entities. For instance, in the alumina-supported Pt, Re, and Pt-Re catalysts, the result of chemisorption shows respective specific areas of 0.29, 0.23, and 0.35m^2/g with corresponding dispersions of 39.1, 31.3, and 48.0%.[5] The Pt-Re catalyst, which has a specific area of 0.35m²/g, consists of 0.26 m²/g from Pt and 0.09 m²/g from Re. But the sum of the specific surface areas of pure components (0.3% Pt or 0.3% Re) is almost 1.6 times higher than Pt-Re catalyst. Also in the Pt-Re catalyst, Re-specific area has decreased in pure Re to one third of its pure state value. This indicates that almost two thirds of the Re has 'alloyed' with Pt.

Since the ability of bulk alloy is not a necessary condition for a system to be of interest as a catalyst, terms such as bimetallic aggregates or bimetallic clusters have recently been adopted in preference to alloys.[6] This refers to highly dispersed bimetallic entities present on the surface of a support such as silica or alumina.

Several other aspects have important effects on the catalyst performance. These include their pore structure (which determines their internal surface area, compressive strength, and density and their size and shape (which determine their geometric surface area per unit particle volume, their bulk or packed density, and the back-pressure characteristics of their assembly). The design of new, more stable catalysts which resist sintering is a constant concern of the catalytic scientists.

Many catalysts are used in the form of impregnated particles. The supported ingredients are generally imposed upon the support from the solution. This introduces many additional factors which must be considered, such as its reactivity with catalyst ingredients, blockage of the pores of the support with catalytic material, and epitaxial effect of the support in altering the crystallinity of the catalytic ingredients added thereonto.

Commercial catalyst preparation technology is so highly proprietary that many crucial details may be missing. In many cases, it is not even patented for the reason that manufacturers keep the process secret. Preparative techniques are, however, beyond the scope of this book. The reader is referred to a recent book describing a detailed account of the typical preparative procedures for the catalysts which are well known to the practicing catalytic scientists.[7]

Chemical reactions can follow quite complex paths and sequences. In the complete form, a mechanism is defined by a change in the rearrangement of atoms, where the exact geometric path that separates reactants from products is specified. In catalysis, in general, one looks for the elementary steps and adsorbed surfaces intermediates; some are stable enough to be isolated. However, in heterogeneous catalysis, a different intermediate may be on each crystal face or only at specific surface defects. The surface structure is heterogeneous on the atomic scale and there are many surface sites that are distinguishable by their number of nearest neighbors.[8] There are surface atoms that can form terraces, steps, and kinks, and these structures may, in turn, contain point defects, adatoms, and vacancies (Figure 1).[8] The structure of the catalyst surface, however, is one of the key factors which controls chemical selectivity.

III. KINETICS AND REACTION MODELING

Mechanism of surface catalysis is very complex. Hougen and Watson[9] and others[10-12] have broken down the steps for fluid-phase reactions that occur on a molecular scale in the following manner:

1. Mass transfer of reactants from the main body of the fluid to the gross exterior surface of the catalyst particle
2. Diffusion of the reactant molecules from the exterior surface of the catalyst particle into the interior pore structure
3. Adsorption of the reactants on the surface
4. Reaction on the surface-molecular rearrangements at active surface sites
5. Desorption of chemically adsorbed species from the surface of the catalyst
6. Diffusion of the products into the fluid

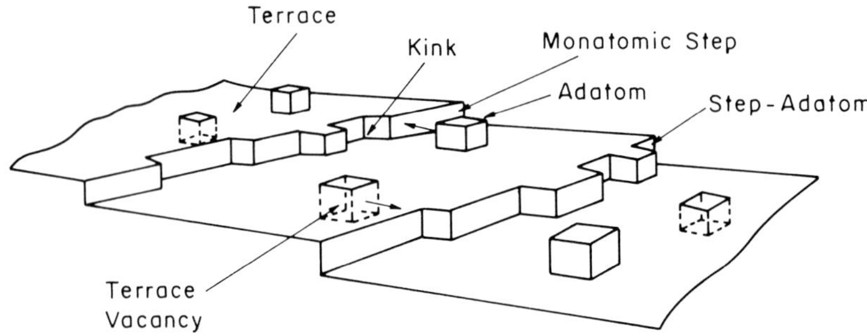

FIGURE 1. Model of heterogeneous solid surface depicting different surface sites. These sites are distinguished by their number of nearest neighbors. (From Spencer, N. D. and Somorjai, G. A., *Rep. Prog. Phys.*, 46, 1, 1983. With permission.)

7. Mass transfer of products from the exterior surface of the particle into the bulk of the fluid

Steps 1, 2, 6, and 7 are physical processes, while steps 3 and 5 are basically chemical in character. Steps 1 and 7 are highly dependent on the fluid-flow characteristics of the system. The mass velocity of the fluid stream, the particle size, and the diffusional characteristics of the various molecular species are the pertinent parameters on which the rates of these steps depend. It has been observed, however, that many industrial catalysts are so active that diffusion is often the controlling factor.

Adsorption is an essential precursor to surface catalysis. Many equations for adsorption equilibria have been advanced on both experimental and theoretical grounds. One of the earliest and simplest is that of the Langmuir, which has been outstandingly successful in the interpretation of adsorption behavior and surface catalysis. However, for certain oxidation reactions, the kinetic rate expressions are derived based on a two-stage redox mechanism, originally suggested by Mars and Van Krevelen.[13] The latter proposes the removal of actual lattice oxygen atoms by direct interaction with the reductant, which must be chemisorbed.[14]

For the Mars-Van Krevelen mechanism, the following terminology is used: a ‘redox component’ is a species or a vacancy required for an elementary step in the redox mechanism; a ‘redox ensemble’ is a set of redox components that can interact with each other; a ‘redox system’ is the combination of redox ensembles required for the restoration of the oxidation activity of the catalyst after one product molecule has been formed.

Certain idealized conditions are taken as the basis for the Langmuir theory. When applied to chemical reactions, this theory is called the Langmuir-Hinshelwood formulation.[15] The following mechanism for reaction on a surface is considered:

1. Reactants adsorb onto the surface
2. Reactants react in adsorbed state
3. Products desorb from the surface

The rate at which these individual reactions occurs is difficult to determine. Therefore, assumption is made that the adsorbed intermediates are in equilibrium with the gas phase, and the overall reaction is proportional to the surface concentrations of adsorbed reactants.

A somewhat more explicit approach was established by Hougen and Watson[9] who derived rate equations in terms of surface concentrations of adsorbed species and free sites and then expressed these concentrations in terms of the Langmuir isotherm. Hougen and Watson were responsible for analyzing a number of possible reaction mechanisms in these terms and for popularizing this approach to analyzing catalytic reaction rate data. For simplicity, such ideal surface models will be referred to as Langmuir-Hinshelwood-Hougen-Watson (LHHW) formulations.

Yang and Hougen[16] have identified several additional reaction mechanisms and have developed tables by comparison with the fundamental form

$$\text{Rate} = \frac{\text{(kinetic terms) (driving force)}}{\text{(adsorption term)}^n} \quad (1)$$

The n is the number of active sites participating in the reaction. The rate equation for each case may be developed using the proper entries from their table.

Although physicochemical insight and several formalisms limit the spectrum of possible LHHW models for a catalytic reaction, a detailed kinetic analysis requires obtaining precise experimental data followed by model discrimination and parameter estimation. Froment[17] has discussed in detail the estimation of parameters and has emphasized the necessity of the statistical testing of the results. Kittrell et al.[18] have shown that substantially different conclusions about the describing model and its parameter values can be drawn, depending upon the type of analysis and data weighting chosen and have stressed the need of nonlinear least squares analysis of the catalytic rate models. Accuracy of the experimental data is no guarantee of a successful kinetic analysis when the experimental program suffers from a lack of insight into the effect of different variables and is poorly designed. Regression analysis can sometimes lead to substantially different conclusions about the describing model and its parameter values due to poor experimental design and lack of physicochemical insight. The estimated parameter values may be strongly interdependent in some cases.

A nonlinear regression algorithm[19] is used to obtain a mathematical fit for various rate expressions derived from the proposed mechanisms. The program minimizes the residual sum of squares (R.S.S.) during the regression.

$$\text{R.S.S.} = [\hat{r}(j) - r(j)]^2 \quad (2)$$

where $\hat{r}(j)$ = calculated rate

and $r(j)$ = experimental rate

The nonlinear computer program basically improves upon the initial estimates of various constants of the rate equation until the R.S.S. can no longer be reduced. Approximately 95% confidence intervals for various constants are calculated from estimates of their individual variances (σ). An estimate of the variances of each model are calculated from the R.S.S. and the appropriate degrees of freedom (DF). The various models are compared by forming a variance ratio. An F test with the proper degrees of freedom is then used to determine if the variances are significantly different at the 95% level. The starting values of the parameters of the various models are estimated by linear regression. The statistical methods of parameter estimation and model selection presented in this section apply equally to redox and power law models.

$$\sigma^2 = \frac{\text{R.S.S.}}{\text{D.F.}} \quad (3)$$

$$F = \frac{\text{lack-of-fit mean sum of squares}}{\text{pure-error mean sum of squares}} \quad (4)$$

The mechanistic models are constructed with the implicit assumption that the underlying model parameters are independent and have some physicochemical significance. The interdependence of the parameter values is inevitable in nonlinear regression, and especially large when a regression variable (or combination of regression variables) is almost constant within a series of experiments. This problem is generally overcome by taking all experimental series (in which different conditions are varied one at a time) and performing a single combined regression.

Rate expressions are eliminated from considerations when any of the converged adsorption equilibrium constants become negative. Models can also be rejected on the basis of improper trends with the temperature, when, in actuality, these adsorption equilibrium constants are positive or have negative temperature coefficients. Of course, physical mechanisms which have such properties as adsorption equilibrium constants with positive temperature coefficients are extremely rare. If the constants of a model are actually unacceptable, the model has generally been rejected. The correlated rate equation is then introduced by substituting the Arrhenius temperature-dependency relation for each of the constants.

Hosten and Froment[20] proposed a method based on the concept that the variance given by Equation 3 is an unbiased estimate of the experimental error variance for the correct model only. For all other models, bias is introduced as a result of lack of fit. A test of the homogeneity of the estimates of the experimental error variance for each model should, therefore, provide an adequate discriminatory criterion. The chi square test can be used for this purpose:

$$\chi^2 = \frac{(\ln s^{-2}) \sum_{i=1}^{m} (DF)_i - \sum_{i=1}^{m} (DF)_i \ln s_i^2}{1 + \frac{1}{3(m-1)} \left[\sum_{i=1}^{m} \frac{1}{(DF)_i} - \frac{1}{\sum_{i=1}^{m} (DF)_i} \right]} \tag{5}$$

where $(DF)_i$ is the degrees of freedom associated with the ith estimate of error variance plus lack of fit s_i^2; s^2 is the pooled estimate of variance plus lack of fit; and m is the number of models.

Well aware of the inherent limitations of any kinetic model, Sharma and Srivastava[21,22] showed that in addition to statistical data interpretation, the resulting rate model must be based on insight into the true physicochemical nature of the process. These authors carried out a series of hydrocarbon and alcohol oxidations over V_2O_5-based catalyst, and showed that the oxidation process is well described by the kinetic equations developed using modified Hinshelwood mechanism as well as by a two-stage redox model, resulting in an identical rate expression. The significance of the parameters of the two models was, however, different. Therefore, from a purely statistical point of view, it was impossible to say that one of these models best described the data. Such a situation requires other experimental evidences, such as performing an *in situ* X-ray diffraction study.[23] These analyses are described in detail in Chapter 3.

IV. CATALYST DEACTIVATION

Catalyst poisoning is one of the least well-understood areas relating to catalysis. To understand poisoning, it is necessary to define carefully the reaction conditions, the catalyst type, and reaction. For example, sulfur is considered to be a severe catalyst poison, but, for example, a nickel catalyst, under certain conditions, is quite tolerant to a concentration of sulfur or sulfur compounds which under lower temperature conditions are severe poisons. Further, certain sulfur compounds, such as H_2S, COS, and CS_2, poison a catalyst, whereas other compounds, for example, thiophene, may very well pass through the catalyst bed and remain undecomposed, thus, not poisoning the catalyst. Much characterization work is needed to provide a more fundamental understanding of poisoning and of introduction of poison tolerance in catalyst systems.

Many commercial catalysts are gradually deactivated by deposition of coke that accompanies the main reaction process. Experimental insights into the manner of poisoning can, of course, be realized by the measurement of the poisoned-to-unpoisoned-rate ratio. As a consequence, the kinetic study of the main reaction in itself becomes seriously complicated.

In spite of several studies made of the kinetics of deactivation processes, there is still no generalized approach to analyze this problem because of the wide variation in the chemical aspects of coke formation.[24-28] Earlier studies contemplated the catalyst activity as dependent only on the reaction time which revealed very little information about the mechanism of deactivation.

It is generally accepted that the rate of main reaction in presence of coking can be expressed as

$$(-r_A) = (-r_A)_0 \cdot a \tag{6}$$

where a is termed as 'activity'. Some authors prefer to use ϕ_A or ψ_A as a deactivation function for activity. The kinetic equation for the main reaction is given by $(-r_A)_0$ and is determined by using kinetic data at zero coke content. It is obvious that some relationship which gives the value of the activity during reaction is required along with Equation 6. It is at this point where the controversies still exist.

Earlier, an equation with separable variables for the deactivation of the form

$$-da/dt = f_1(T) \cdot f_2(C_i) \cdot f_3(a) \tag{7}$$

was used.[24] The overall chemical reaction rate expression consisted of a reaction-rate function multiplied by a deactivation function as

$$(-r_A) = f_4(T) \cdot f_5(C_i) \cdot a \tag{8}$$

Thus, by keeping the reaction mixture composition and the temperature constant, the variation on the rate of reaction due solely to the deactivation process can be studied directly. This method of analysis of experimental data is valid only in the case where concentration-dependency function is previously evaluated on the basis of experiments and which are generally of power law type.

The above method of analysis was later extended by several authors[27,28] to the point where the kinetics of the main reaction was of the LHHW type or at which the variables were nonseparable

$$-da/dt = \psi(p_i, T) \cdot f_3(a) \tag{9}$$

where ψ (p_i, T) is the deactivation function.

Corella and Asua[29] generalized theoretically the kinetics of deactivation by coking where Equation 9 was accomplished and obtained the equation:

$$-da/dt = \psi(p_i, T) \cdot a^d \tag{10}$$

where $d = (m + h - 1)/m$, m and h are the number of active sites involved in the controlling step of the main reaction and the deactivation reaction, respectively. This method has been successfully applied in several studies[30,31] of catalyst deactivation by coke formation, and is considered to be a more fundamental approach to the analysis of catalyst deactivation.

V. STRUCTURE OF THIS BOOK

A few concluding remarks are appropriate in order to place in context the basis of the contents of the present book. The opportunities for understanding catalysis have improved sharply over the last 2 decades since a number of sophisticated analytical instruments have been developed for the study of surfaces, the complex interfaces at which catalysis takes place. The instrumental techniques are being applied to practical catalysts and are providing new insight into catalytic behavior. The availability of such techniques has opened up possibilities of a clearer understanding of surface phenomena and therefore of the fundamental nature of catalysis.

Advances in understanding the structure and chemistry of catalysts do not result from the application of any one experimental technique, but rather from the application of the specific combination of techniques best suited to the system under study. The required understanding and the successful design of new catalysts and their exploitation demand the integration of solid state physics, inorganic chemistry, organic chemistry, physical chemistry, surface chemistry and physics, and chemical reaction engineering, including fluid mechanics, heat, and mass transfer.

Given the key importance of surface characterization and physicochemical kinetics, this book is structured to review the modern techniques of characterization and to treat several "case histories" of a specific industrial class of catalysts, each concerned with an industrial process or class of processes which will serve to illustrate the rapidly accelerating need of catalytic science over recent years. The kinetics and modeling will be introduced as they arise in the context.

REFERENCES

1. **Larrabee, G. B. and Shaffner, T. J.,** *Anal. Chem.*, 53, 163R, 1981.
2. **Werner, H. W.,** *Electron Microsc.*, 3, 200, 1980.
3. **Solymosi, F.,** *Catal. Rev.*, 1, 233, 1968.
4. **Benson, J. E. and Boudart, M.,** *J. Catal.*, 4, 704, 1965.
5. **Jothimurugesan, K., Nayak, A. K., Mehta, G. K., Rai, K. N., Bhatia, S., and Srivastava, R. D.,** *AIChE J.*, 31, 1997, 1985.
6. **Sinfelt, J. H.,** *Bimetallic Catalysts*, John Wiley & Sons, New York, 1983.
7. **Stiles, A. B.,** *Catalyst Manufacture*, Marcel Dekker, New York, 1983.
8. **Spencer, N. D. and Somorjai, G. A.,** *Rep. Prog. Phys.*, 46, 1, 1983.
9. **Hougen, O. A. and Watson, K. M.,** *Chemical Process Principles*, (Part 3), John Wiley & Sons, New York, 1947.
10. **Carberry, J. J.,** *Chemical and Catalytic Reaction Engineering*, McGraw-Hill, New York, 1976.
11. **Hill, C. G.,** *An Introduction to Chemical Engineering Kinetics and Reactor Design*, John Wiley & Sons, New York, 1977.
12. **Bischoff, K. B. and Froment, G. F.,** *Chemical Reactor Analysis and Design*, John Wiley & Sons, New York, 1979.
13. **Mars, P. and Van Krevelen, D. W.,** *Chem. Eng. Sci.*, 3, 41, 1954.
14. **Agarwal, D. C., Nigam, P. C., and Srivastava, R. D.,** *J. Catal.*, 55, 1, 1978.
15. **Hinshelwood, C. N.,** *The Kinetics of Chemical Change*, Clarendon, Oxford, 1940.
16. **Yang, K. H. and Hougen, O. A.,** *Chem. Eng. Prog.*, 46, 146, 1950.
17. **Froment, G. F.,** *AIChE J.*, 21, 1041, 1975.
18. **Kittrell, J. R., Hunter, W. G., and Watson, C. C.,** *AIChE J.*, 11, 1051, 1965.

19. **Draper, N. R. and Smith, H.,** *Applied Regression Analysis,* John Wiley & Sons, New York, 1966.
20. **Hosten, L. H. and Froment, G. F.,** *Proc. 2nd Int. Symp. Chem. Reaction Eng.,* Dechema, Heidelberg, 1976.
21. **Sharma, R. K. and Srivastava, R. D.,** *AIChE J.,* 27, 41, 1981.
22. **Sharma, R. K. and Srivastava, R. D.,** *AIChE J.,* 28, 855, 1982.
23. **Srivastava, R. D., Stiles, A., and Jones, G. A.,** *J. Catal.,* 77, 192, 1982.
24. **Khang, S. J. and Levenspiel, O.,** *Ind. Eng. Chem. Fundam.,* 12, 185, 1973.
25. **Hegedus, L. and Petersen, E. E.,** *Catal. Rev.,* 9, 245, 1974.
26. **Butt, J. B.,** *Chemical Reaction Engineering, Adv. Chem. Ser. 109,* American Chemical Society, Washington, D.C., 1972, 259.
27. **Corella, J., Asua, J. M., and Bilbao, J.,** *Can. J. Chem. Eng.,* 59, 647, 1981.
28. **Froment, G. F.,** *Proc. 6th Int. Cong. Catalysis,* Bonds, G. C., Wells, P. B., and Thompkins, F. C., Eds., Chemical Society, London, 1977.
29. **Corella, J. and Asua, J. M.,** *Ind. Eng. Chem. Process Des. Dev.,* 21, 55, 1982.
30. **Srivastava, R. D. and Guha, A. K.,** *J. Catal.,* 91, 254, 1985.
31. **Pal, A. K., Bhowmick, M., and Srivastava, R. D.,** *Ind. Eng. Chem. Process Des. Dev.,* 25, 236, 1986.

Chapter 2

EXPERIMENTAL METHODS IN CATALYTIC RESEARCH

I. TECHNIQUES FOR SURFACE CHARACTERIZATION

In the fields of heterogeneous catalytic science and research, the appreciation and interpretation of the characteristics of catalysts and their modification for a particular purpose requires detailed characterization of solids and their surfaces. However, the characterization of surfaces and surface products are problematic and challenging.

In the past, most of the techniques utilized for the physicochemical characterization of catalyst surfaces were intended for use with the highly dispersed systems. *In situ* analysis by infrared spectroscopy,[1-4] Mössbauer spectroscopy,[5] electron spin resonance,[6] and X-ray diffraction[7] are frequently carried out.

The two most commonly used diffraction techniques for bulk analysis are X-ray diffraction (XRD) and transmission electron microscopy (TEM), and are now considered as 'classical'. TEM is useful in obtaining information about the size and distribution of the catalyst particles along with the crystal structure of active material and support, and also detailed chemical composition at a microscale. Lattice parameters are commonly obtained from X-ray diffraction measurements. XRD measurements can also offer the dimensions of small crystallites. Scherrer formula is commonly used for determining the average dimension of roughly equiaxed particles.[8]

A major change in the practice of catalysis in the past 2 decades has been the concerted application of surface analytical techniques to catalysts. Surface techniques can be characterized by their capability for providing information on a micrometer scale. Transmission electron microscopy can be considered as the first method available for the determination of physical structures of surface topography on a nanometer scale. Since then, a large number of surface analytical methods have gradually been introduced and improved. Recently, the most commonly used methods for thin-film and surface analysis have been critically reviewed and described by Werner and Garten.[9] This excellent compilation has emphasized the need for a synergetic multimethod approach in thin-film analysis.

The analysis of both the physical and chemical structures of surfaces is performed primarily by means of photons, electrons, ions, and neutral atoms. A number of interaction processes between exciting beam and solid matter is used for surface analytical methods, forming their initial working base. An updated list of the characterization techniques used in catalysis is given in Table 1.[10] A description of the most widely used (ESCA, AES, and IRS) and of some less frequently used of these methods, but relatively new to catalysis (Mössbauer, EXAFS, PIXE, and RBS), will be given in the next section. Numerous excellent reviews of the remaining techniques have appeared recently[9] and are summarized in Table 2. Some of these techniques are limited to single-crystal studies. They will be briefly presented here.

In this book, it has not been possible to treat any techniques exhaustively and some have had to be omitted entirely. These shortcomings can be remedied in part by reference to pertinent publications listed in each section. However, the selected guides to the literature concerning physicochemical characterization in general and some applications to catalysis in particular are offered in Table 3.

A. X-Ray Photoelectron Spectroscopy

X-ray photoelectron spectroscopy (XPS) or ESCA (electron spectroscopy for chemical analysis) is a technique for measuring electron-binding energies. As a result, information can be obtained on dispersant/support interactions and on subtle surface changes.

Table 1
COMMON METHODS FOR SURFACE CHARACTERIZATION

Surface Analysis Method	Acronym	Physical Basis
Auger electron spectroscopy	AES	Electron emission from surface atoms excited by electron, X-ray, or ion bombardment
X-ray and ultraviolet photoelectron spectroscopy	XPS, UPS, (ESCA)	Electron emission from atoms
Infrared spectroscopy	IRS	Vibrational excitation of surface atoms by absorption of infrared radiation
Ion-scattering spectroscopy	ISS	Elastic reflection of inert gas ions
Low-energy electron diffraction	LEED	Elastic backscattering of low-energy electrons
High-resolution electron energy loss spectroscopy	HREELS	Vibrational excitation of surface atoms by inelastic reflection of low-energy electrons
Secondary-ion mass spectroscopy	SIMS	Ion-beam induced ejection of surface atoms as positive and negative ions
Extended X-ray absorption fine structure analysis	EXAFS	Interference effects in photoemitted electron wave function in X-ray absorption
Proton-induced X-ray emission	PIXE	Excitation of inner-shell electrons by protons
Rutherford back-scattering spectrometry	RBS	Elastic back-scattering of ions by using He^+ beam
Mössbauer effect spectroscopy	MES	Hyperfine interactions
Thermal desorption spectroscopy	TDS	Thermally induced desorption or decomposition of adsorbates

Table 2
SELECTED REVIEWS RELATED TO LEED, HREELS, IRS AND RAMAN, ISS, AND SIMS TECHNIQUES

Author	Year	Special topics	Ref.
Koestener et al.	1983	LEED, HREELS	11
Marcus and Jona	1982	LEED for surface structure	12
Hair	1967	IRS in surface chemistry	1
Poate and Buck	1976	ISS: theory and application	13
Baun	1976	ISS: applications	14
Buck	1977	ISS: experimental	15
Benninghoven et al.	1982	SIMS	16
Long	1977	Raman	82

Photoelectrons are ejected from the sample by monoenergetic X-ray excitation (about 1 keV), $h\nu$. The kinetic energy (KE) of the ejected electrons is analyzed, and the peaks in the resulting kinetic energy spectrum correspond to electrons of specific binding energies (B.E.) in the sample:

$$\text{B.E.} = h\nu - \text{KE} \tag{1}$$

The binding energy of the photoelectron is the core-level energy relative to Fermi level. These binding energies are dependent on the kind of atom, its valence state, its environment, and the penetration depth. The electron spectral lines are referred to by the symbol of the element and the indication of the atomic energy levels involved in the emission process. The quantum numbers $n = 1,2,3,\ldots..$, $l = s, p, d, \ldots$, and $J = l + s$ are used for photoelectron lines.

Table 3
FIELDS OF APPLICATION: REVIEWS RELATED TO CATALYSIS

Author	Year	Special topics	Ref.
Park	1976	Comprehensive survey of methods	17
Briggs	1979	ESCA	18
Ponec	1980	Electron spectroscopy	19
Thomas and Lambert	1980	General characterization	20
Haller	1981	Vibrational spectroscopic applications	21
Mc Guire and Halloway	1981	AES: applications	22
Bhasin	1981	ESCA & AES: applications	23
Zhdan	1982	ESCA and UPS: adsorption and surface reaction	24
Hercules and Klein	1982	ESCA: applications	25
Bonzel	1983	Applications	26
Koestner et al.	1983	Hydrocarbons, UPS, LEED, and EXAFS	11
Backx et al.	1980	HREELS: applications	84
Delannay	1984	General characterization	27

Shifts in the binding energy can be measured and are related to the chemical environment of the atom. Studies of the satellite structure of the spectra, measurement of the relative intensities, as well as the energies of the photoelectron peaks, and determination of the angular distribution of the ejected photoelectron have all added to the characterization of chemical species and have been extensively reviewed.[28-32]

For surface studies, methods are needed for the evaluation of all the elements present as a function of surface depth. Since XPS only gives information over the outermost layers of a solid compound, i.e., its surface layers (15 to 20 Å), this makes it a very useful technique for studying catalytic surfaces. The calibration of the energy scale of the spectrometer is achieved by referencing to an internal standard line. Generally, all binding energies are charge corrected to the C 1s line, assuming the C 1s line for adventitious hydrocarbon to be at 284.6 eV. This carbon is always present as a result of oil contamination from the vacuum system.

When comparing ESCA lines from the same element in different chemical surroundings, shifts in the peak positions can be observed. Chemical shifts arise from the fact that an electron being ejected from an inner atomic shell probes the chemical environment as it leaves the molecules or solid. In simplest terms, the more highly oxidized a given species, the lower the electron density in the valence shell, and consequently, the greater the energy needed to remove a core electron.[33-36]

The theoretical interpretation of core-level chemical shift has been attempted at various levels of sophistication. They range from total energy difference calculations to rough electrostatic models:[37] One of the primary reasons why chemical shifts can be analyzed by such a variety of methods is that their origin is so simply and directly connected to the molecular orbital charge distribution. In turn, it is very often this charge distribution that is of primary interest in a given investigation.[38]

ESCA spectra are typically presented as the energy distribution of the emitted electrons by plotting intensity or beam current vs. the energy. Figure 1 shows the Ni2p spectra for the four $NiO\text{-}Al_2O_3$ catalysts, together with those of pure specimens of nickel oxide and nickel aluminate used as standards. For the catalyst samples, the shake-up peaks and $2p_{3/2}$ peak are identical (Table 4). The standards, however, show marked differences, the most prominent being the double feature of the $2p_{3/2}$ peak for NiO. The binding energies and main satellite structures of the catalysts are typical of several likely Ni^{2+} species, including NiO and $NiAl_2O_4$. Comparison with the Ni2p spectrum for Ni^o on Al_2O_3 revealed conclusively that no Ni^o was present. The binding energy of the $2p_{3/2}$ peak for the $NiAl_2O_4$ standard is

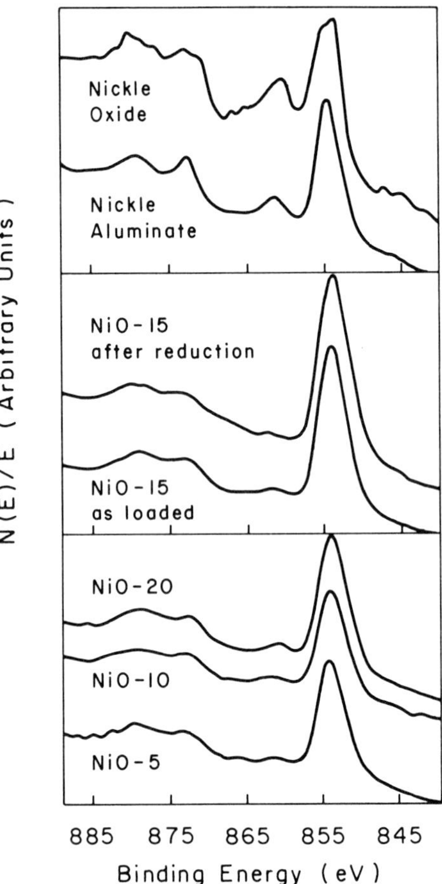

FIGURE 1. Ni 2p spectra for the supported catalysts and NiO and nickel aluminate standards. (Reprinted with permission from Srivastava, R. D., Onuferko, J., Schultz, J. M., Jones, G. A., Rai, K. N., and Athappan, R., *I & E C Fundam.*, 21, 457, Copyright 1982, American Chemical Society.)

Table 4
BINDING ENERGIES

	$Ni2p_{3/2}$	O1s	Al2p
NiO — 5	854.8	530.8	73.8
NiO — 10	854.8	530.8	73.9
NiO — 15	854.7	530.7	73.8
NiO — 15 (reduced)	854.4	530.6	73.8
NiO — 20	854.7	530.8	73.7
$NiAl_2O_4$	855.2	530.6	73.8
NiO	854.2	529.7	
	855.6	531.1	

Reprinted with permission from Srivastava, R. D., Onuferko, J., Schultz, J. M., Jones, G. A., Rai, K. N., and Athappan, R., *I & E C Fundam.*, 21, 457, Copyright 1982, American Chemical Society.

0.4 higher than those of the catalyst specimens. However, matrix effects and multiple species would well account for this relatively small displacement.

From the intensity of the peaks, a quantitative composition of the surface layers can be deduced. Methods for quantifying the XPS measurements utilizing peak-area sensitive factors and peak-height sensitivity factors have been developed. Ratios based on peak areas are a more reliable source of information on which to base atomic concentrations. This is specifically true for transition metals, where the occurrence of prominent shake-up is frequent.[39]

B. Auger Electron Spectroscopy

Auger electron spectroscopy (AES) is a surface-sensitive analytic technique which allows the elemental composition in the surface or near-surface region to be determined. AES is a secondary process. When a core level of a surface atom is ionized by an impinging electron beam (1 to 10 keV), the atom may decay to a lower energy state through an electronic rearrangement which leaves the atom in a doubly ionized state. The energy difference between these two states is given to the ejected Auger electron which will have a kinetic energy characteristic of the parent atom. For example, the energy of KL_1L_3, Auger transition following ionization of the core, K, level is the difference of the L_1 and L_3 energies from that of the K level. The spectroscopic notation n = K, L, M, . . . , and 2j = I, II, III, . . . is commonly used in Auger electron notation. When the Auger transitions occur within a few angstroms of the surface (escape depths are approximately 10 to 20 Å), the Auger electrons may be ejected from the surface without loss of energy and give rise to peaks in the secondary electron energy distribution function. The energy and shape of these Auger features can be used to unambiguously identify the composition of the solid surface. The most prominent Auger transitions observed in AES are given in Figure 2. All elements above helium produce Auger peaks in the 0 to 2000 eV range.

The basic components required for AES are an ultrahigh vacuum system, an electron gun, and an energy analyzer for detection of Auger electron peaks in the total secondary electron energy distribution. A high-sensitivity cylindrical mirror analyzer is the most widely used energy analyzer in AES. Auger currents of the order of $10^{-12} < I < 10^{-14}$ Å are obtained in this way. Because the Auger peaks are superimposed on a rather large continuous background, they are most easily detected by differentiating the energy distribution function N(E). Thus, the conventional Auger electron is the function dN(E)/dE — $d[I(E_2)]/dE$. Electronic differentiation is accomplished by superimposing a small a.c. modulation voltage to the slowly changing energy-selecting voltage applied to the cylindrical mirror analyzer and feeding the modulated output of the electron multiplier to a phase-sensitive lock-in amplifier.

The reason for signal differentiation is made apparent as illustrated in Figure 3. The curve labeled N(E) depicts the standard energy distribution for electrons emitted from a silver target with the incident beam of 1 keV. The large peak at 1000 eV is attributable to the elastically reflected electrons. The N(E) $\times$ 10 curve gives an indication of the Auger transitions. Since this represents a small signal on a larger varying background, the advantages of signal differentiation via modulation, and the use of phase-sensitive detection to improve the signal-to-noise ratio are clear.

The peak-to-peak magnitude of an Auger peak in a differentiated spectrum generally is directly related to the surface concentration of the element which produces the Auger electrons. Quantitative analysis may be accomplished with varying degrees of accuracy by comparing the peak heights obtained from an unknown specimen with those from pure elemental standards or from compounds of known composition. Seah[40] proposed the use of the negative peak-to-background height rather than the customarily and most widely used Auger peak-to-peak height. However, other workers and reviewers[9] recommend the determination of the area under the $I(E_2)$ peak to obtain a measure of the Auger current in practical work.

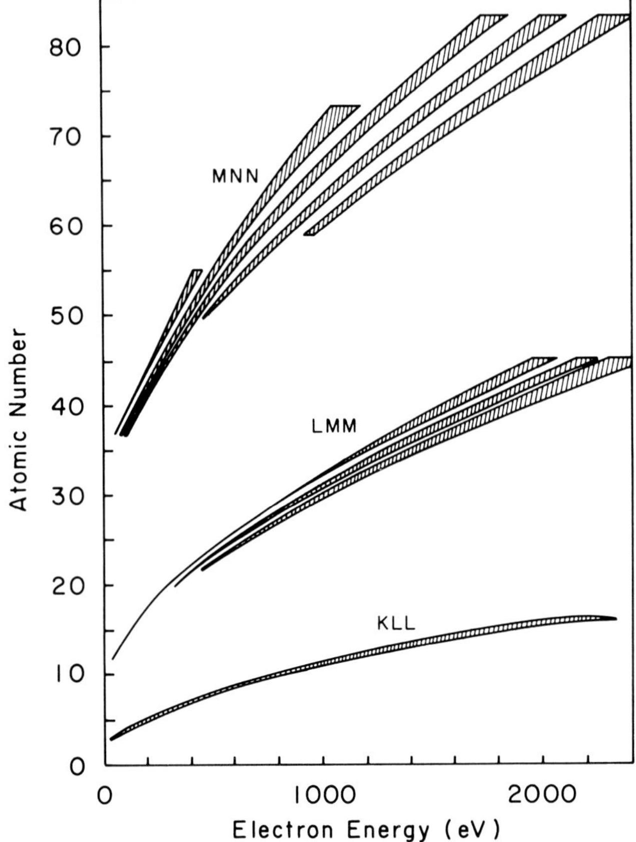

FIGURE 2. Most prominent Auger transitions observed in AES. (From Davis, L. E., MacDonald, N. C., Palmberg, P. W., Riach, G. E., and Weber, R. E., *Handbook of Auger Electron Spectroscopy,* Perkin-Elmer, Eden Prairie, Minn., 1978. With permission.)

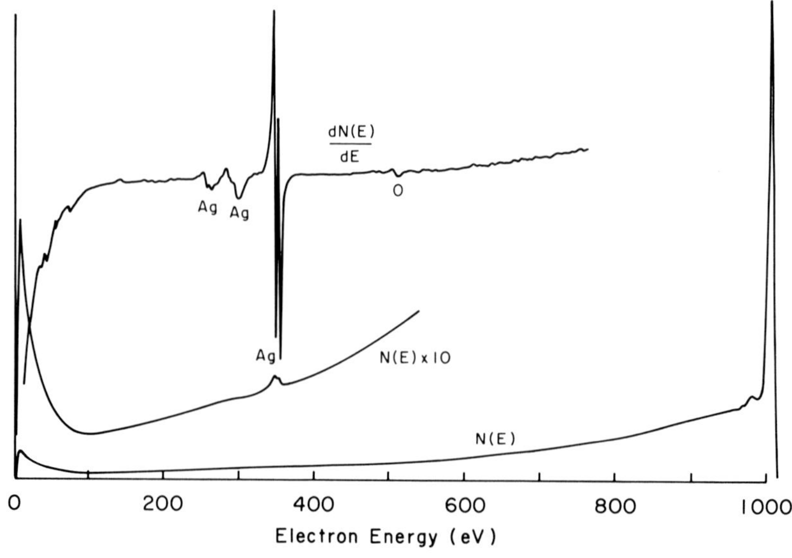

FIGURE 3. Comparison of AES peak display for silver.

A more detailed discussion on AES and its ramifications can be found in several reviews of this technique.[22,41-45]

C. Infrared and Raman Spectroscopy

Infrared spectroscopy (IRS) is a proven valuable technique in catalyst research. Since it monitors molecular vibrations, this technique is very useful to changes in molecular environment and structure. It has been routinely used in catalyst characterization.

Today, an IR experiment is generally performed using a Fourier transform infrared spectrometer (FTIR). This employs an interferometer rather than the prism or grating monochromator used in dispersive instruments. The optical assembly consists of three parts: a source, an interferometer (generally a Michelson interferometer), and a detector. Light from the source is collimated, then split into two beams of equal intensity by a beam splitter. A beam in each arm of the interferometer reflects the light onto the beam splitter, where it is recombined and passed through the sample chamber to the detector.[79]

Interferometers produce interferograms rather than wavelength spectra. It is a recording of the intensity that results from the constructive and destructive interference for all the wavelengths of light output from the interferometer as a function of pathlength difference between the two arms of the interferometer. The interferogram is converted to a wavelength spectrum by means of a Fourier transform.

FTIR can be used to study many diverse types of catalysts and catalytic systems. This also allows determination of structure by providing data indicative not only of ligands such as CO and NO, but ligands like hydrocarbons. The very rapid scanning capability allows many transient experiments to characterize catalytic reactions, ligand exchanges, and even agglomeration of metal. The development and refinement of new IR sampling techniques should facilitate and augment *in situ* studies of functioning catalysts.[80]

Complementary to IR is Raman spectroscopy, in that the molecular vibrations that produce strong IR transitions generally produce weak Raman transitions and vice versa. For example, the molecular vibrations of high surface-area oxides and oxide supports have strong IR bands, but very weak Raman bands. On the other hand, the transition-metal oxides supported on high surface-area oxides have strong Raman transitions and weak IR transitions.

A Raman spectrum is a measure of the intensity of light scattered inelastically from a molecule. The frequency of a Raman active vibrational transition is the frequency difference between the incident and scattered light. Only one set of optics is required to record the 10 to 4000 cm^{-1} range of vibrational transitions, since Raman spectra are measured in the visible wavelength region. This technique is less sensitive to sample morphology. The instrumentation and sampling techniques have been developed to the extent where the number of applications of Raman spectroscopy to the study of catalysts and catalytic processes is beginning to rival that of IR spectroscopy.

The theoretical and experimental aspects of IR and Raman spectroscopy are well covered by Alpert et al.,[81] Long,[82] and Griffiths.[83] No further details will be presented here.

D. Extended X-Ray Absorption Fine Structure

Extended X-ray absorption fine structure (EXAFS), which determines the interatomic distances to, coordination numbers of, and relative disorders of several coordination shells of atoms surrounding the absorbing atom, has recently been applied to the study of supported metal catalysts.[46-51] It can provide uniquely new electronic and structural information about supported metals and metal-support interactions. Some excellent reviews have recently been published.[51-56]

EXAFS involves, in addition to determining the position of absorption edge, analysis of the fine structure which exists immediately above the X-ray absorption edge. The fine structure can be Fourier-inverted to provide quantitative structural information about the immediate environment around each element. Curve fitting can circumvent some of the limitations of the Fourier transform inversion of the data and represents a potentially powerful analysis technique. Each element present can be examined separately because the absorption edges of each are sufficiently separated that the fine structures do not overlap. Comparison of the radial structure data obtained for each element present provides quantitative information about the structure. The location of the absorption edge provides information about the effective charge on each element. The intensity of the transition to bound d states provides a measure of the extent of filling of the d orbitals, a quantity of intense interest to catalytic scientists.

The theoretical basis for EXAFS analysis,[57-63] based on the scattering of ejected photoelectrons by atoms in the coordination shells surrounding the central absorbing atom, is for spherically symmetric, essentially infinite geometry. If the photon energy is sufficiently greater than the core-level binding energy, the photoelectron can be considered as essentially free. The ejected electron (photoelectron) is characterized by a wave factor K, which is then related to its energy, $E - E_0$, by

$$h^2 K^2/2m = E - E_0 \tag{2}$$

where E and E_0 are the X-ray photon energy and the energy of the absorption edge. The X-ray photon energy, E, equals hw, where w is the photon frequency in radians per second determined by dividing the speed of light by the wavelength of the photon. The wavelength of the spherically outgoing electron wave, $\lambda_e = 2\pi/K$, is related to energy of the electron by the above relation. The constants, h and m, are Planck's constant divided by 2π and the mass of the electron, respectively.

The fine structure, X(K), is found by the following operation:

$$X(K) = [\mu(K) - \mu_0(K)]/\mu_0(K) \tag{3}$$

The function, $\mu(K)$, is the total absorption coefficient observed experimentally. The term, $\mu_0(K)$, is the monotonically decreasing part of the absorption coefficient from which any pre-edge absorption contributions have been subtracted.

The cross-section for absorption of the X-ray photon depends on the interaction between the wave function of the initial core state and the final state of the electron. The initial state is fixed and does not vary with the frequency of the photon. For the case of an isolated atom, the photoelectron would exist only as an outgoing state and no fine structure would be observed. If the outgoing wave backscattered from surrounding atoms, the resulting ingoing waves could interfere either constructively or destructively with the initial state at the origin to alter the final state wave function. Since the phase between the ingoing and outgoing waves and their interaction depends upon the photon frequency, one observes an oscillatory behavior in the absorption spectrum. The phase of the final state is modified also by the shift, $\alpha_1(K)$, due to the absorbing atom potential and a phase shift, ϕ (K), associated with the backscattering amplitude. The period of the EXAFS oscillations in K space depends upon the distances of the backscattering atoms from the central absorbing atom.

The EXAFS single-scattering theory is

$$X(K) = \Sigma A_j(K) \sin[2 K R_j + 2 \partial_j(K)] \tag{4}$$

The factor $A_j(K)$ is an amplitude function for the jth shell:

$$A_j(K) = \frac{N_j}{K R_j^2} F_j(K) \exp(-2 K^2 \sigma_j^2) \tag{5}$$

These expressions sum the different contributions to the fine structure of the jth coordination shell of N_j atoms at an equilibrium distance, R_j. The contribution from each shell is proportional to the number of atoms in that shell. The term $F_j(K)$ is a factor accounting for electron backscattering and inelastic scattering.[64] The exponential is the Debye-Waller broadening term, which smears out the EXAFS oscillations through thermal and structural disorder. The phase shift, $\alpha_j(K)$, is the sum of the absorbing atom and backscattering atom phase shifts. The root mean-square deviation, σ_j^2 from R_j, is assumed small compared to interatomic distances.

If the phase shift follows a linear function of K such that

$$\partial_j(K) = a_0 + a_1K \tag{6}$$

then

$$KR_j + a_0\, a_1K = a_0 + (a_1 + R_j)K \tag{7}$$

and the linear part of the phase shift occupies an equivalent position in the sine term as an interatomic distance.

The Fourier transform of Equation 4 leads a function ϕ_n (R):

$$\Phi_n(R) = (1/2\pi)^{1/2} \int_{K_{min}}^{K_{max}} K^n\, X(K) \exp(-2i\, K\, R)\, dK \tag{8}$$

where R is the distance from the absorber atom. This radial distribution function has peaks centered about interatomic distances, which differ from the real distances by an amount about equal to the linear part of the phase shift shown in Equations 5 and 7. The integer, n, takes on value from 1 to 3 depending upon the atomic number of the backscattering atoms.[65] The limits of the transform are from K(min) > 3Å, to a maximum value of K for which the signal-to-noise ratio is acceptable. A value of n = 3 has been found to give a radial distribution that is reasonably insensitive to a choice of both K(min) and the energy corresponding to K = 0. Cutoff errors in K space will cause a beating in the Fourier transform. This beating can be reduced by taking K(min) and K(max) to be nodes in the data.[51]

Two types of Fourier transforms can be used. One can use theoretical values of the phase shift and transform with respect to exponential part to obtain a function that will peak around the value of R_j[65]. The alternative is to use the transform of Equation 8, which peaks at $R_j + a_1$ as indicated by Equation 7. This shift of magnitude a_1 should be transferable from a standard whose interatomic distance is known and whose atomic environment is similar. In the treatment of EXAFS data, it is useful to obtain an inverse Fourier transform of the function $\phi_n(R)$ over a limited range of R. This procedure determines the contribution to EXAFS arising from shells of atoms within that range of R. In the analysis of EXAFS data on bimetallic clusters, two EXAFS functions are considered, one for each component of the clusters.[66]

As an example, Figure 4 shows the EXAFS data taken at liquid nitrogen temperature for a platinum on titania containing 1.7% platinum after H_2 reduction of the samples at the indicated temperature.[50] The EXAFS function X(K) is plotted against the photoelectron wave factor (K) after the smooth background decay has been subtracted from the data above the absorption edge (Figure 4a). The associated Fourier transforms of the functions, which were taken over from 2.2 to 11.0 $Å^{-1}$ in K-space, are shown in Figure 4b. The main peak of each transform was isolated by Fourier filtering below 1.0 Å and above 3.5Å and was back-

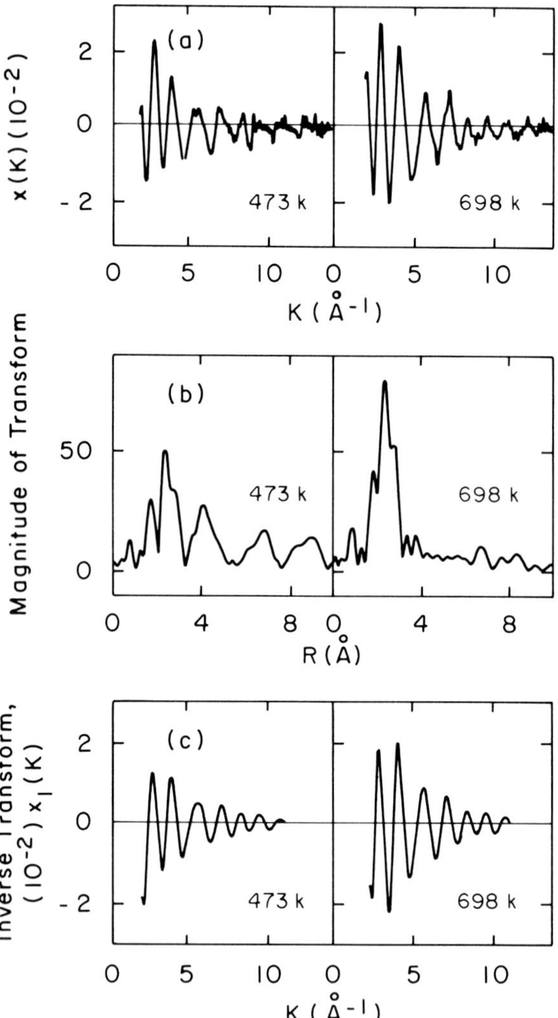

FIGURE 4. (a) Normalized EXAFS data at the indicated temperatures; (b) Fourier transforms of K^3-weighted EXAFS; (c) Inverse transforms of the first coordination shell (From Short, D. R., Mansour, A. N., Cook, J. W., Sayers, D. E., and Katzer, J. R., *J. Catal.*, 82, 299, 1983. With permission.)

transformed to K space. The resulting inverse transforms are shown in Figure 4c. The subscript 1 in $X^1(K)$ signifies that the function includes only the contribution to EXAFS arising from the first coordination shell of metal atoms.

The EXAFS data for the 473 K H_2 reduction is significantly smaller in relation to the other sample. This has been attributed to an increase in the average first-shell coordination number. Also, there is a slight shift in peak position toward shorter R with increasing reduction temperature (Figure 4b); both the peak distance and the shape of the fine structure envelopes in Figure 4b suggested that the main peak for each curve corresponds to a first coordination shell consisting of Pt atoms. The inverse data correspond to a coordination shell composed of only one Pt-Pt interaction.

E. Proton-Induced X-Ray Emission

Proton-induced X-ray emission (PIXE), which represents a highly sensitive method of elemental method of analysis, has come to be established as an important analytical tool in the study of the near-surface elemental composition of supported metal catalysts.[67-69] A beam of energetic protons ($1 < E_p < 4$ MeV) with beam spots of 1 μm $< \Delta x <$ 10 mm in diameter is used for the excitation of inner-shell electrons and the resulting characteristic X-rays are used for the identification and quantification of the element of interest in the sample. α-Particles from radioactive sources can also be used for excitation in X-ray emission analysis.[70] In the case of supported metal catalysts, the metal atoms will be distributed over several microns in depth. In such cases, PIXE can give the in-depth concentration of the metal.

Although the use of characteristic X-rays of the elements has also been the basis of electron beam microanalysis and X-ray fluorescence, the use of protons and other heavily charged particles for excitation is a recent development.[71] It has several advantages over electron beam microanalysis and X-ray fluorescence. In the case of electron beam excitation, the major contribution to the background arises from the (bremsstrahlung) generated by the deceleration of these electrons, while in the case of proton excitation the background arises due to the scatter of these photons by the electrons of the target material. The bremsstrahlung background from the protons, however, is very small. The major factor which limits the lower detection limit in PIXE is the nature of the analyte. In the case of thick insulating targets, the secondary electrons bremsstrahlung can contribute substantially to the background in the low-energy region, thereby reducing the sensitivity for the lighter elements. The secondary electron bremsstrahlung can be minimized by neutralizing the positive charge build-up by spraying electrons from an electrically heated carbon filament onto the target during proton bombardment.[72]

The proton beam is obtained from a Van de Graff accelerator. The accelerated protons are momentum-analyzed by a magnet and focused using a quadrupole doublet onto a diffuser. The beam becomes diffused by mulitple scattering. The central homogeneous portion of the transmitted beam is normally selected using a set of collimators and is incident on the target (catalyst). The X-rays produced are detected with a lithium-drifted silicon, Si(Li), detector.

The absolute concentration of the elements can be calculated provided the proton energy, the X-ray production cross-section, and the detector efficiency as a function of X-ray energy are known. A simpler and more accurate approach is to calibrate the system for a fixed beam-target-detector geometry using standard thin targets in the form of number counts per microgram of the element per microcoulomb of the proton charge is shown in Figure 5. A similar approach for thick targets may not always be feasible. However, as long as only the ratio of elements in near-surface regions and its variation due to changes in experimental conditions is required, one can still use the overall efficiency factor for the thin targets in this case also.

A typical PIXE spectra of the calcined and reduced Pt-Re-Al_2O_3 (0.3 wt% each) catalyst is shown in Figure 6. The X-ray lines have been identified and labeled. This figure reveals that Re concentration seem to predominate over Pt in the reduced state.

F. Rutherford Backscattering Spectrometry

Rutherford backscattering spectrometry (RBS) is one of the most powerful microanalytical techniques available for the study of the absolute surface concentrations as well as the depth distributions. Being based on the elastic backscattering of the ions, the method is truly nondestructive. It is highly sensitive, surface concentrations of the order of monoatomic layer being amenable for study when present on a lighter matrix.[73,74]

It was recently pointed out that this was a useful technique for analyzing thin films.[70,75] This technique was designed originally for research in nuclear physics, and tended to be temperamental and complicated. Most accelerators are now designed for material analysis; however, the full potential of the RBS technique in catalyst research has not yet been exploited.

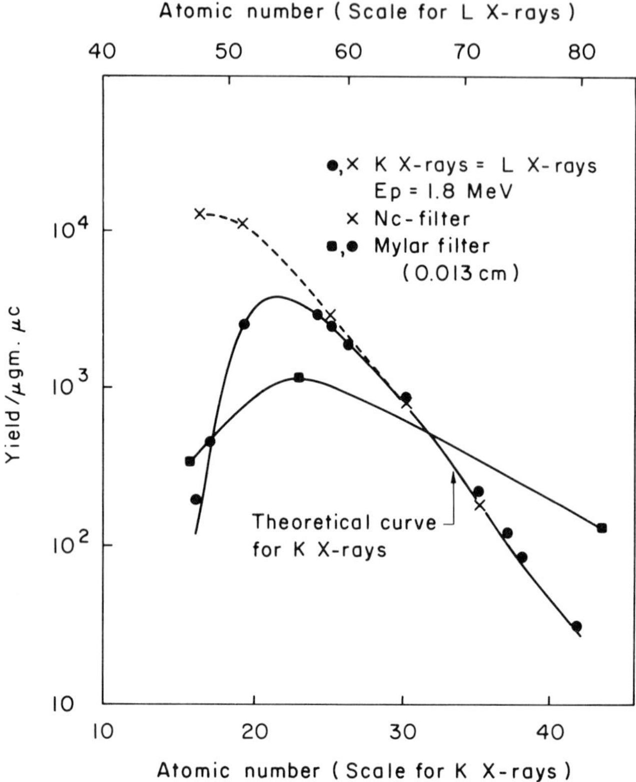

FIGURE 5. X-ray detection efficiency of a PIXE system as a function of atomic number for 1.8 MeV protons.

In the case of supported metal catalyst, at high temperatures, significant redistribution of the metal atoms in the support material can take place due to diffusion. RBS can be very useful in studying this problem. Further, the fact that the support is always a light matrix like alumina, silica, etc., is an additional attraction. This technique has been recently applied for supported monometallic model systems[77] and bimetallic catalysts.[69,77]

1. Principles of RBS

The kinetic energy, E, of a particle of mass M_1 having an initial kinetic energy E_0 after an elastic collision with a stationary particle of mass M_2 is given by

$$E = KE^0 \tag{9}$$

where K is the kinematic factor which can be derived from conservation of energy and momentum. In the laboratory coordinate frame

$$K = \left[\frac{M_1 \cos\theta + (M_2^2 - M_1^2 \sin^2\theta)^{1/2}}{M_1 + M_2}\right]^2 \tag{10}$$

θ being the scattering angle.

When the scattering nucleus is buried inside the target, the projectile will lose an energy ΔE_1 to the medium before reaching the scattering nucleus and the scattered projectile will lose an energy ΔE_2 in emerging through the target. The kinetic energy of the backscattered projectile as it emerges out of the target is given by

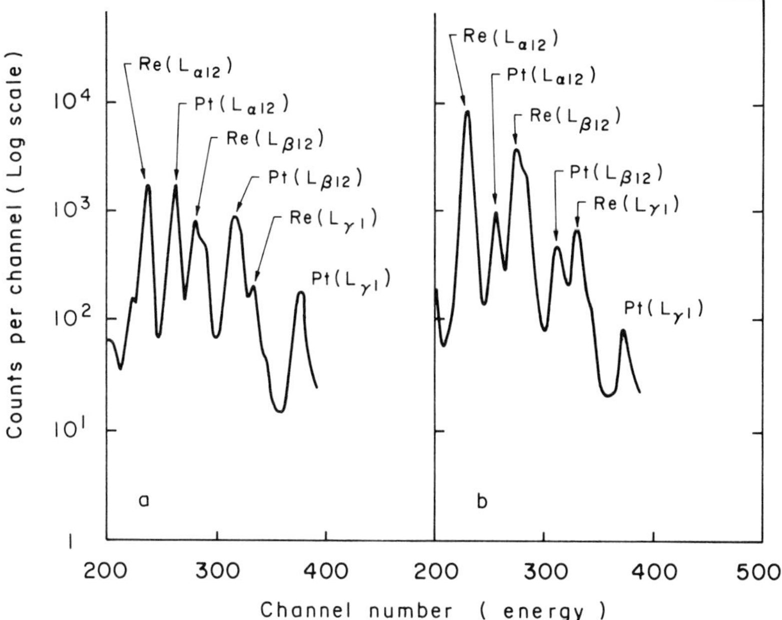

FIGURE 6. PIXE spectrum of Pt-Re-Al_2O_3 catalyst: (a) calcined, (b) reduced. (From Jothimurugesan, K., Nayak, A. K., Mehta, G. K., Rai, K. N., Bhatia, S., and Srivastava, R. D., *AIChE J.*, 31, 1997, 1985. Reproduced by permission of the American Institute of Chemical Engineers.)

$$E = K(E_0 - \Delta E_1) - \Delta E_2 \tag{11}$$

The measured shift ($K\,\Delta E_1 + \Delta E_2$) in the energy of the scattered projectile from KE_0 is a measure of the depth at which the scattering nucleus is situated from the surface. The conversion from energy scale to actual thickness can be achieved by knowing the stopping cross-section, density, and composition of the target-material.

From Equation 9, it can be shown that the best mass resolution is obtained for $\theta = 180°$. However, to place a detector exactly at 180° with respect to the primary beam is not possible. The detector is normally positioned at some steeply backward angle such as 170°, and hence the name backscattering spectrometry. With annular detectors, scattering angles very near to 180° can be reached. It is noted that for backscattering to occur, M_1 should be less than M_2. He^+ are the most widely used projectiles in backscattering spectrometry. While the mass resolution can be improved by increasing the projectile energy, it is also necessary to ensure that no nuclear reactions take place. Energies in the range 1 to 3 MeV for He^+ and 300 to 500 keV for H^+ are generally used.

While the kinematic factor contains 'mass' signature of the scattering nucleus, to obtain the concentration it is necessary to know how frequently such collisions occur. This information is contained in the differential scattering cross-section. The differential cross-section for an elastic scattering can be calculated using the principles of conservation momentum and kinetic energy, and a specific model for the force that acts during collision between the projectile and the target. For most cases, the force is represented by the coulomb repulsion of the two nuclei, as long as the distance of closest approach is large compared to nuclear dimensions, but small compared to atomic dimensions. With these assumptions, the differential scattering cross-section is given by Rutherford's formula:[74]

$$\frac{d\sigma}{d\Omega} = \left[\frac{Z_1Z_2e^2}{4\ E}\right]^2 \frac{4}{\sin^4\theta} \frac{\{[1 - ((M_1/M_2)\sin\theta)^2]^{1/2} + \cos\theta\}^2}{[1 - ((M_1/M_2)\sin\theta)^2]^{1/2}} \tag{12}$$

where

- Z_1 = atomic number of the projectile
- Z_2 = atomic number of the target
- σ = Rutherford backscattering cross-section
- Ω = solid angle subtended by the detector

From Equation 12 it is clear that because the backscattering cross-section is directly proportional to Z_2^2, the backscattering spectrometry is highly sensitive for heavier elements rather to lighter elements.

2. Depth Profile Calculation

When the beam particles pass through a finite-thick, monoisotopic target, they lose some of their energy. As a consequence, the particles backscattered from the rear surface of the thick film have less energy when they are detected than particles backscattered at the front surface. Scattering events that take place somewhere between front and rear surfaces are recorded at some intermediate energies. If the beam is unattenuated, the scattering probability at any depth is proportional to the number of atoms of a particular kind present there. In this way a concentration profile of a given element is translated into a signal of corresponding height and decreasing energy in the backscattering spectrum. The energy loss of the particles along their inward and outward paths can be calculated from the stopping cross-section, ϵ. The detected energy difference between particles scattered from the front and back surfaces of the film is given by the product of Nt and ϵ, where N is the atomic density and t, the film thickness. However, the energy-to-depth conversion is direct if the atomic density of the sample is known, and linear if the energy loss is independent of incident energy. In the case of thick and compound films, suitable approximations are made.

The stopping cross-section can be mathematically given as

$$\epsilon = 1/N\ dE/dx \tag{13}$$

where dE/dx is specific energy loss. The kinematic factor is then $K = E_1/E_0$, where E_1 is the energy of the outcoming beam and E_0 that of the ingoing beam.

If E is the energy of the beam immediately before scattering at a depth x, as illustrated in Figure 7, then $E_0 - E$ is the energy loss along the inward path ΔE_{in}; similarly, $KE - E_1$ is the energy loss along the outward path ΔE_{out}. Energy loss factor S is given by

$$[S] = \left[\frac{K}{\cos\theta_1}\frac{dE}{dx}\bigg|_{in} + \frac{1}{\cos\theta_2}\frac{dE}{dx}\bigg|_{out}\right] \tag{14}$$

Stopping cross-section factor, ϵ, is given as

$$[\epsilon] = \left[\frac{K}{\cos\theta_1}\epsilon_{in} + \frac{1}{\cos\theta_2}\epsilon_{out}\right] \tag{15}$$

Also,

$$E = \Delta[\epsilon]\ Nx \tag{16}$$

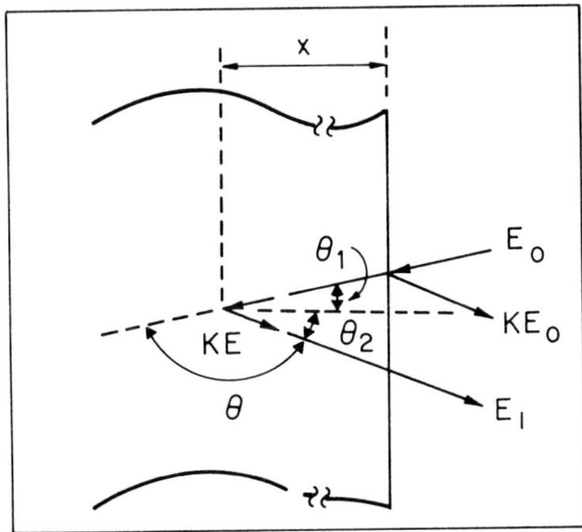

FIGURE 7. Schematic representation of scattering.

$[dE/dx]_{in}$ and $[dE/dx]_{out}$, or ϵ_{in} and ϵ_{out}, can be found from either surface energy approximation or mean energy approximation.

If the depth of penetration is considered to be out into thin slabs, then ${}_nE_1$ is the energy of a particle emerging after a collision in the n slab. Such a depth profile for the first slab is represented by the expression:

$$1E_1 = K_1E - \frac{\Delta x}{\cos\theta_2} \frac{dE}{dx}\bigg|_{K_1E} \qquad (17)$$

with the addition of appropriate terms for inner slabs.

For data analysis it is convenient to have the relation between energy and depth changed to that between height vs. channel number. For a general case, height, H of the spectrum at its channel is $H_i \approx \sigma (E_i)\, Nx/\cos\theta_1$, where σ is scattering cross-section.

3. Compound Sample

The treatment of a compound sample, either a mixture or a chemical compound, differs from that of the monoisotopic case in two significant ways. First, as the probe particles penetrate the film, they lose energy as the result of interactions with more than one element. Consequently, ϵ depends on the composition of the sample. Second, when the probing particles with energy E are scattered at a specific depth within the sample, the value of K and scattering cross-section will depend on the particular mass of the atom they strike. Since the stopping cross-section varies with energy, the energy that the particles lose along identical outward tracks also depends on the mass of the atom struck in the scattering collision. For a compound sample, the yield of the backscattering spectrum and energy-to-depth conversion, thus, depend on the element struck in the collision.

The final expression for the ratio of heights of the spectrum of the sample composed of two monoisotopic elements A and B in the mixture AB, after applying appropriate approximations to ϵ and S, is

$$\frac{H_A^{AB}}{H_B^{AB}} = \frac{N_A^{AB}}{N_B^{AB}} \frac{\sigma_A}{\sigma_B} \frac{[\epsilon]_B^{AB}}{[\epsilon]_A^{AB}} \qquad (18)$$

where H_A^{AB} is the height of the energy spectrum of the particles scattered from atom A in a mixture of AB at the surface. The atomic concentration of A and B in the mixture is N_A^{AB}

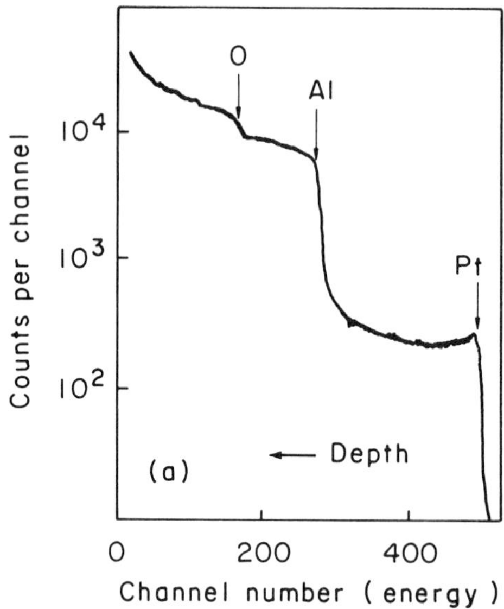

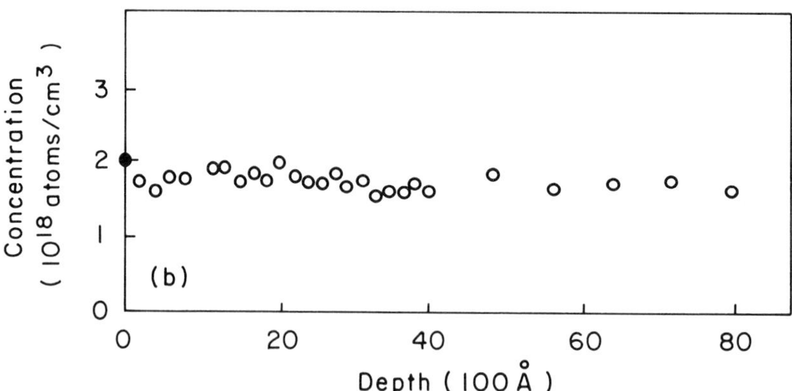

FIGURE 8. (a) Backscattering spectrum of $Pt\text{-}Al_2O_3$ catalyst; (b) depth profile of $Pt\text{-}Al_2O_3$ catalyst.

and N_B^{AB}, respectively. The yield ratio of two signal heights is approximately given by the ratio of the scattering cross-section, which is proportional to $(Z_A/Z_B)^2$. The equation of the signal height becomes

$$\frac{H_A^{AB}}{H_B^{AB}} = \left[\frac{Z_A}{Z_B}\right]^2 \frac{[\epsilon]_B^{AB}}{[\epsilon]_A^{AB}} \tag{19}$$

A computer code is normally developed to calculate the number density by doing height ratio.[78] A typical RBS spectrum for a reduced $Pt(0.3\%)\text{-}Al_2O_3$ catalyst is given in Figure 8a. The corresponding depth profile is shown in Figure 8b.

G. Mössbauer Spectroscopy

Mössbauer emission spectroscopy (MES) is a resonant technique employing a photon source, an absorber, and a photon detector.[85,86] The gamma-ray energy from a radioactive nucleus is modulated by imparting a Doppler velocity to the source. The motion of the source can be achieved in two ways, either constant velocity or constant acceleration. The gamma rays of discrete energies can be resonantly absorbed by absorber nuclei. The number of transmitted photons are plotted vs. photon energy and a peak is observed where resonance occurs. It can be applied to catalysts in which virtually all of the metal atoms are surface atoms.[66,87]

The Mössbauer effect can be observed only for a small number of nuclides. In most cases, this effect is observed in the range 10 keV $< E_0 <$ 150 keV, depending upon the Debye temperature of the element involved. One of the nuclides in which the Mössbauer effect is often observed is ^{57}Fe. Besides this, the ^{95}Pt, ^{119}Sn, ^{121}Sb, and ^{151}Eu nuclides are also sensitive probes.

1. Hyperfine Interactions

The nature of the Mössbauer spectrum and the number and position of the peaks observed in the spectrum are sensitive to the extranuclear environment, which results in different spectra. This is attributed to so-called hyperfine interactions. These include the interaction between the charge distributions of the Mössbauer nucleus and the extranuclear electric and magnetic fields and may be represented by the Hamiltonian

$$\mathcal{H}_0 = E_0 + M_1 + E_2 + \dots \tag{20}$$

where E_0 describes the electric monopole interaction between the nucleus and the surrounding electrons, M_1 represents the magnetic dipole interaction between the nuclear magnetic dipole moment and the surrounding extranuclear magnetic field and E_2 represents the electric quadrupole interaction between the nuclear electric quadrupole moment and the electric field gradient of the surrounding electrons. The terms higher than E_2 are small and can be neglected.

The first term, E_0, causes a change in the energy separation between the ground and the excited states of the nucleus. Depending on the chemical state of the compound, the amount of this change will be different for the absorber and the source. As a result, the position of the peak in the Mössbauer line is shifted and the amount of shift is known as isomer shift (IS) or chemical shift. The presence of the other two terms M_1 and E_2, either singly or in conjunction, causes a splitting of the energy levels and this gives rise to multiple lines in the Mössbauer spectra. The peak positions and intensities of these lines provide valuable information about the hyperfine interactions.

2. Isomer Shift

Isomer shift is one of the important Mössbauer parameters which results from the coulombic interaction between the charge distribution of the nucleus and the electron density at the nucleus (s electrons). As a result of such interaction, the nuclear energy levels do not undergo splitting but are manifested through a slight shift of the Mössbauer energy levels in a compound with respect to that in the free atom. Thus, IS measures the difference between the Mössbauer transition energy in the source and that in the absorber when both of them are at the same temperature and can be expressed as

$$IS = (2/5)\, Z\, e^2[|U_s(O)|^2_A - |U_s(O)|^2_s]\, [R^2_e - R^2_g] \tag{21}$$

where suffixes A and s refer to the absorber and the source, respectively, $U_s(O)$ is the s-electron density at the site of the nucleus and $R_e{}^2$ and $R_g{}^2$ are the mean-squared nuclear radii in the excited and ground state, respectively (Figure 9).

3. Quadrupole Splitting

The above discussion of isomer shift was based on the assumption that the nucleus is

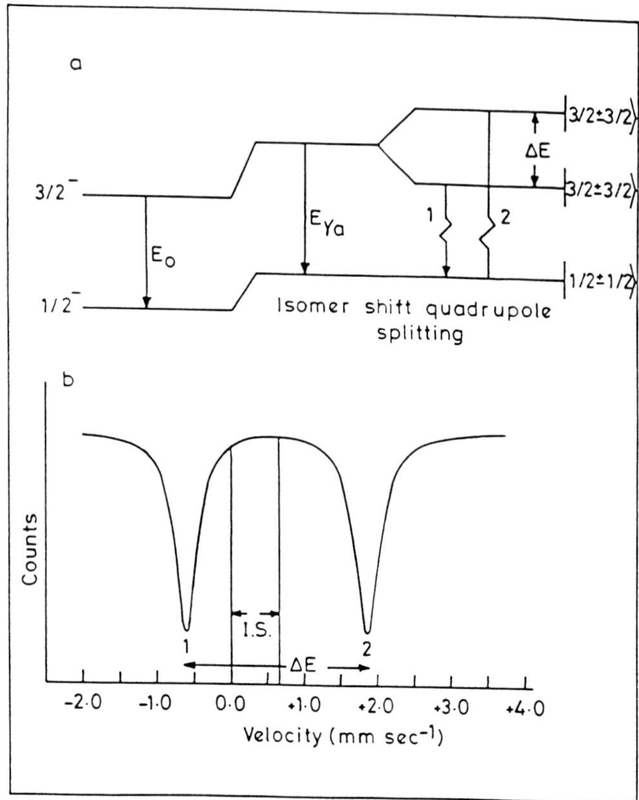

FIGURE 9. Effect of quadrupole interaction on the nuclear energy levels of source/absorber and the resulting Mössbauer spectrogram showing the quadrupole splitting and isomer shift.

spherical and has uniform charge density. A nucleus with spin $>1/2$ has a nonspherical charge distribution and hence, there will be an interaction of the nuclear charge density with the extranuclear electric field. More precisely, the Z-components of the electric field gradient (EFG) tensor will interact with the electric quadrupole moment of the nucleus. Such an electric quadrupole interaction leads to a splitting of the nuclear energy levels. ^{57}Fe nucleus, for example, has spins in the excited and ground state as $I_e = 3/2$ and $I_g = 1/2$, respectively. The higher energy level, I_e, splits into two levels while the I_g remains unsplit (or degenerate) as shown in Figure 9.

Analysis of quadrupole splitting values and isomer shift measurements can provide the oxidation or valence state as well as the coordination number (e.g., iron). Similarly, a study of the temperature variation of quadrupole splitting and isomer shift can be used to investigate temperature-induced structural or chemical changes.

4. Magnetic Hyperfine Splitting

The magnetic dipole moment μ_N of a nucleus having spin I is given by

$$\mu_N = g_N B_N I \tag{22}$$

where g_N is nuclear Landau factor and B_N is Bohr magneton. As a result of the magnetic interaction between this magnetic dipole moment and the applied or local (extranuclear) magnetic fields at the nucleus, the degeneracy of the nuclear energy levels is completely

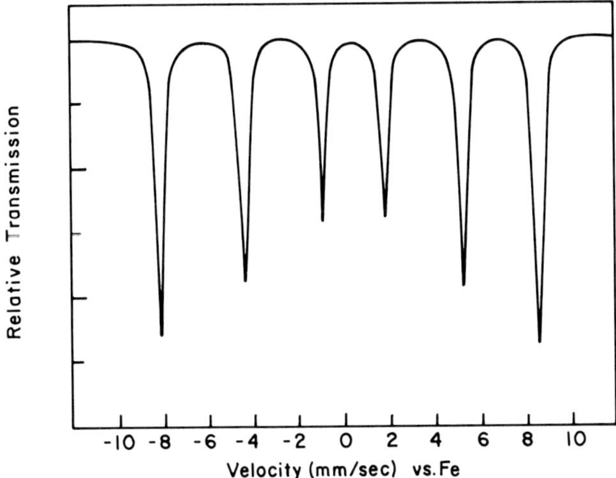

FIGURE 10. The magnetic dipole and electric quadrupole hyperfine structure of α-Fe_2O_3.

removed. Taking into account the different g_N values of the ground and excited states and the selection rules, it can be shown that a symmetric six-line Mössbauer spectrum will be observed. The isomer shift is determined by the center of gravity of the six peaks. Although the linewidths of the six peaks are in general equal, their intensities are always different. For many magnetic substances, the electric quadrupole and magnetic dipole interactions occur concurrently and give rise to a complicated spectra. One of the important applications of the magnetic hyperfine splittings is for the calibration of the Mössbauer spectra (Figure 10).

Mössbauer spectroscopy has been used to characterize the bimetallic reforming catalysts.[66] In addition, it has been successfully applied to Co-Mo/Al_2O_3 catalysts to explain their activity for hydrodesulfurization.[131,132] The formation and subsequent disproportionation of a new phase, CoMoS, accounts for the observed maximum in catalytic activity at intermediate values of α, the atomic ratio of Co/(Co + Mo).

H. Other UHV Surface Techniques

In many cases, diagnostic tools such as ion-scattering spectrometry (ISS), secondary-ion mass spectrometry (SIMS), low-energy electron diffraction (LEED) and high-resolution electron energy loss spectroscopy (HREELS) are incorporated together into the vacuum chamber. They are usually complemented by XPS, AES, and thermal desorption spectroscopy (TDS), which have often proved necessary to obtain an unambiguous interpretation of the spectra. Simple model catalysts, usually based upon a metal single crystal, are prepared and characterized in ultrahigh vacuum (UHV) with respect to their surface structure. In this way, a variety of surface structures and compositions can be tested.

In the following pages these techniques (ISS, SIMS, LEED, HREELS, and TDS) are briefly described as they specifically apply to the single-crystal catalyst studies.

1. Ion-Scattering Spectroscopy

Ion-scattering spectroscopy (ISS) is able to determine the composition of only the outer one or at most two monolayers of a solid. Therefore, information is obtained from both the composition and the short-range structure of the surface.

Briefly, the technique consists of bombarding a target with a beam of noble gas ions (He^+ or Ne^+ are generally used), of energy in the keV range.[13,88-90] The energy spectrum of the

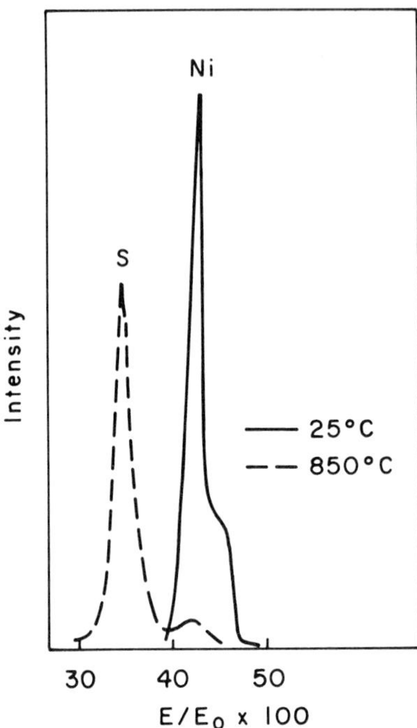

FIGURE 11. ISS spectra for a nickel(100) surface before (-) and after (- -) sulfur adsorption. (From Brongersma, H. H., *J. Vac. Sci. Technol.*, 11, 231, 1974. With permission.)

scattered ions at a particular scattering angle (usually 90°) is recorded while keeping the energy of the incident ions fixed. The energy of the scattered ions is then characteristic for the mass of the target atom.

In particular, if E_0 is the kinetic energy of a noble gas ion of mass m_1 and E_1 is its kinetic energy after being scattered at an angle of 90° during an elastic collision with a surface atom of mass m_2, the following relation holds.[13]

$$E_1/E_0 = (m_2 - m_1)/(m_2 + m_1) \tag{23}$$

Thus, by selection of the primary ion m_1 and measuring the scattered ion energy E_1/E_0, m_2 can be calculated and the element can be identified. Figure 11 illustrates the ISS spectra (Ne^+ ion impact) for the (100) face of nickel before and after sulfur adsorption.[88] The adsorbed structure was obtained by heating a weakly sulfur-doped nickel crystal. The small nickel peak in the spectrum for the heat-treated crystal is due to an incomplete surface monolayer.

Recently, Kelley and co-workers[91-93] have reviewed its use for analyzing metal surfaces and also reported its application to highly dispersed supported Pt catalysts. ISS has also been successfully applied to study the direct and rapid determination of the atomic ratio of the metal in the topmost layer in surface alloys.[94] The limited mass resolution is a major problem in ISS.

2. *Secondary-Ion Mass Spectrometry (SIMS)*

In SIMS, a beam of energetic ions (500 to 15,000 eV), typically Ar, is used to sputter atoms, molecules, and cluster from a solid surface placed in a vacuum. A fraction of these particles is emitted in a charged state. These secondary ions emitted from a small volume of the sample are separated according to their mass-to-charge ratio by means of a mass spectrometer. Therefore, other than high-vacuum system, only two critical components need to be acquired in order to perform a SIMS analysis. These are the ion source and a mass spectrometer for detecting secondary ions.

Among a variety of modes of SIMS application, two variations are currently in use in catalysis. In the first, the primary beam is focused to a diameter of approximately 100 nm, allowing high spatial resolution. When the ion beam actually removes layers of material from the surface (typically 1 $\mu m\ h^{-1}$), the operation is referred to as 'dynamic'. This mode results in in-depth analysis.[9]

The static SIMS mode uses an extremely low primary-ion beam density (10^{-9} Å cm^{-2}) and a low primary-ion energy (0.5 to 2 keV) resulting in a negligible sputtering rate.[16] In this way, a nearly undisturbed surface of the topmost monolayer can be studied. This requires UHV conditions to avoid contamination and reaction with the residual gas components.

The use of SIMS for catalyst research is unique, since the secondary ions are believed to arise only from the top one or two layers of the sample, surface sensitivity is high when compared to other surface analysis methods such as ESCA and AES. Furthermore, a fraction of the secondary ions appears as molecular fragments, which are often characteristic of the surface chemistry. Also, the sensitivity of certain elements is very high; ppm analyses of monolayers may be possible.

In many cases, SIMS provides unique information on the chemistry and structure of a surface. It has been successfully applied to probe the local atomic structure of the solid catalyst surfaces[95] including the zeolites[96] as well as to distinguish differences in the degree of metal support interaction in catalysts.[97,98]

In single-crystal studies, SIMS measurements have yielded kinetic and structural information for small gas molecules reacting on single-crystal surfaces.[99-101] As an example, Figure 12 shows the variation of the NH_4^+, NH_3^+, NH_2^+, and NH^+ ion yields with temperature in the decomposition of ammonia over an Fe(110) single-crystal sample.[101] This classical work reveals that above 350 K, these species decompose to N_{ads} and H_{ads}, which ultimately desorb at higher temperature, presumably to form NH_3.

There are, however, many disadvantages associated with SIMS. The two weak points are: (1) calibration of SIMS for quantitative work using sensitive factors is often hampered by the lack of a suitable standard sample with a close match to the sample to be measured and (2) the cluster formation processes may be quite complicated, making an interpretation of results equivocal.

Thus, Benninghoven et al.[16] concluded that the current state-of-the-art in SIMS 'has succeeded in developing an excellent and powerful analytical technique, but based upon processes that are not yet sufficiently understood'.

3. *Low-Energy Electron Diffraction*

Low-energy electron diffraction (LEED) has become an extremely useful tool for basic research on catalysis.[19] LEED is used in establishing whether or not substrates and adsorbed species are regular in their geometric arrangements. In particular, this technique determines the bond distances and the angles of the adsorbed hydrocarbons when they are ordered into an overlayer lattice. This method, like XRD, provides the quantitative description of the molecular structure present on the surface.

LEED makes use of the surface sensitivity of low-energy electrons. The angular distribution of electrons of energies (10 to 500 eV) elastically diffracted from a sample is measured. Only elastically reflected electrons are detected by the use of a high-pass filter. This is achieved by employing several grids. Since such slow electrons can traverse only very short

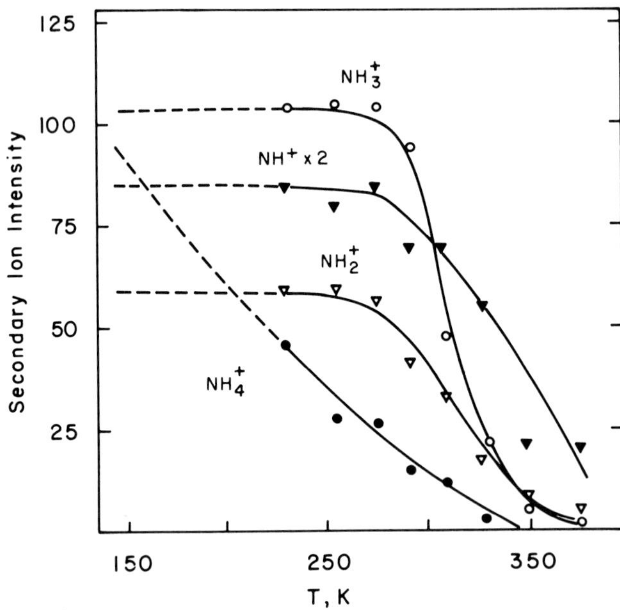

FIGURE 12. SIMS intensity change of NH_3^+ during slow heating of an Fe(110) face saturated with $NH_{3,ad}$ at low temperature. (From Dreschler, M., Heinkes, H., Kaarmann, H., Wilsch, H., Ertl, G., and Weiss, M., *Appl. Surf. Sci.*, 3, 217, 1979. With permission.)

mean-free paths into as well as out of a solid sample, this technique is highly selective to surfaces. UHV better than 10^{-9} torr, preferably in the 10^{-12} region, is required.

The intensity I_{hk} of the diffracted beam is determined by the arrangement and the core electron density of the atoms, i.e., by the elemental composition. The diffracted electron intensity is measured as a function of the electron energy; a set of intensity vs. voltage.

LEED patterns could be used as a reliable measure of the coverage. For example, LEED studies have been effectively carried out for the adsorption of oxygen on silver crystals,[102] when applying a suitable calibration procedure. A range of oxygen coverages can be produced by direct exposure to oxygen. An (n × 1) LEED pattern indicates an oxygen atom coverage of 1/n monolayers, where one monolayer is taken as the number of silver atoms in the troughs on the (110) surface (8.4×10^{14} cm^{-2}). One Langmuir is equal to an exposure at 10^{-6} torr for 1 sec. LEED must not, however, be considered as a 'ready-to-use' technique for quickly giving quantitative results on arbitrary surface samples.[9]

4. High-Resolution Electron Energy Loss Spectroscopy

High-resolution electron energy loss spectroscopy (HREELS), though not quantitatively, is probably the most versatile surface structural tool available today; it is a technique for the detection of vibrations of atoms and molecules adsorbed on single-crystal metal surfaces.

Like LEED, this technique also makes use of the surface sensitivity of low-energy electrons (2 to 10 eV). These electrons are strongly damped in the crystal lattice, so the backscattering fraction that emerges from the sample should carry only surface information to a depth of about three atomic layers. About 1 out of 10^3 reflected may lose discrete amounts of energy by excitation of vibrating dipoles in the form of adsorbed molecules, the inelastically scattered electrons being packed closed to the specular direction.[84]

There are two variations in the scattering mechanism. In the first, referred to as dipole approximation, these inelastic diffraction processes take place in two steps: an inelastic

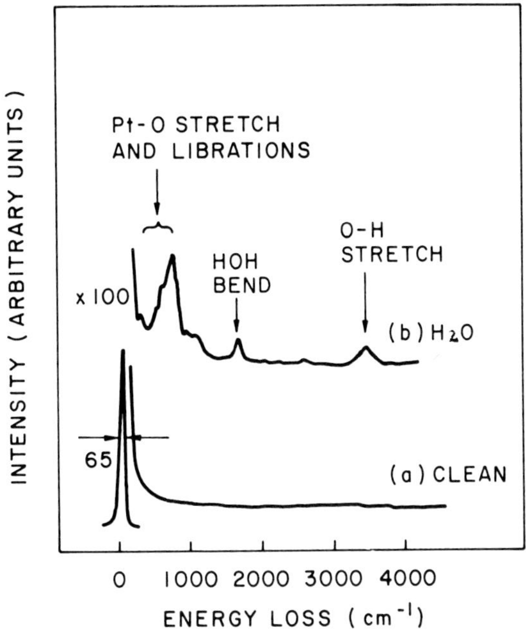

FIGURE 13. HREEL spectra of (a) clean Pt(111) surface; (b) Pt (111) after exposure to H_2O at 100 K. (From Fisher, G. B. and Sexton, B. A., *Phys. Rev. Lett.*, 44, 683, 1980. With permission.)

reflection followed or preceded by excitation of the vibrating dipoles. The second scattering mechanism may also give significant contribution to the intensity of loss features. This mode is referred as 'impact scattering'. A review of both experimental and theoretical research on these states is given by McRae.[103]

The dipole approximation[104,105] has been shown to correctly describe the intensities of energy loss peaks as a function of electron impact energy and results in realistic charges of adsorbed species.[106,107] By measuring the absolute intensity of a loss peak as a function of the angle with respect to the direction of the specularly reflected beam, the contributions from dipole and impact scattering may be separated.[82]

The basic components required for HREELS are an ultrahigh vacuum system, an electron-beam generator, an electron spectrometer for energy analysis of the backscattered electrons, and a detection system. For focusing the beam, deflectors, either 180° hemispherical or 127° cylindrical, are generally used in the assembly of HREELS spectrometers. A spectral range from 30 (240 cm^{-1}) to 500 meV (4000 cm^{-1}) in 10 min by a single reflection from a single-crystal surface may be covered. The HREEL spectrum[108] in Figure 13 obtained for low-temperature monolayer H_2O adsorption on Pt(111) surface shows the presence of two main vibrations characteristic of water: the HOH scissor mode (h-bonding) at 1630 cm^{-1} and the OH stretch at 3420 cm^{-1}. In addition, vibrations assigned as Pt-O stretch and librations are also visible.

HREELS has been used successfully to provide information on the specific adsorption sites and geometrics of adsorbed molecules, the state of hybridization of hydrocarbons, and surface reactions,[84,108-113] as well as the structural effects on the reactivity of surfaces by varying the exposed crystal plane.[114]

5. *Thermal Desorption Spectroscopy*

Thermal desorption spectroscopy (TDS) is widely applied to the study of adsorption-

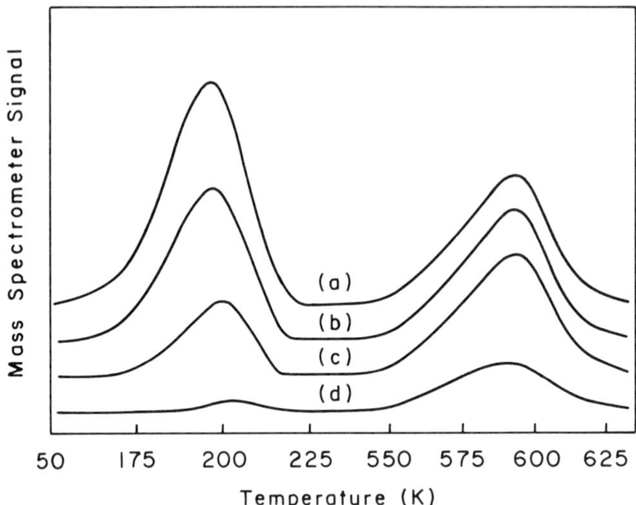

FIGURE 14. Desorption spectrum from molecularly adsorbed and atomically adsorbed oxygen from Ag(110) following exposure at 128 K: (a) 600L; (b) 180L; (c) 60L; (d) 15L. (From Barteau, M. A. and Madix, R. J., *Surf. Sci.*, 97, 101, 1980. With permission.)

desorption phenomena on single-crystal substrates.[115] This is usually carried out in an ultrahigh vacuum system. On-line real-time analysis with a mass spectrometer is generally carried out. Typically, desorption signals of the order of 10^{-8} torr are measured. UHV better than 10^{-10} torr having nominal pumping speeds of approximately 10^5 cm^3 sec^{-1} are required. A principal advantage of thermal desorption methods is that the desorbing product and reactant gases do not readsorb.

Sample heating by radiation functions well. A programmable (current mode) 400- to 1000-W DC power supply is generally adequate for normal rates of 1 $ksec^{-1}$. Cooling is achieved via conduction using a liquid nitrogen reservoir. The amount of the adsorbed gas is determined from the area under the thermal desorption spectrum. The coverage, N (molecules/cm^2), is calculated from the following relationship:

$$N = Z \times F \times N_A \times A/(a \times S) \tag{24}$$

where F is the system-pumping speed; N_A is Avogadro's number expressed in molecules $torr^{-1}$ $liter^{-1}$; A is the area (amp seconds) under the flash peak; a is the dosed crystal area; S is the sensitivity of the mass spectrometer (amp $torr^{-1}$) to the particular mass of interest. Z is the ratio of the areas of an indirectly flashed desorption peak to that of a direct flash.

Analysis of TPD data is illustrated in Figure 14 for oxygen desorption on Ag(110) for varying exposure.[116] Two features of the desorption spectra are evident: oxygen adsorbs at 120 K into a molecular state which desorbs at 200 K, and an atomic state which desorbs at 600 K. The relative magnitude of the low and high temperature desorption states are sensitive to the exposure. TDS has also been applied to very small crystals of the zeolite[117] as well as to high-area catalysts.[118]

Other temperature-programmed thermal methods, temperature-programmed reaction spectroscopy (TPRx) and temperature-programmed surface reaction (TPSR), are also useful in catalyst research. These two methods also utilize a thermal transient, in the form of a linear temperature vs. time curve. The response, evolution of product gas (TPSR), indicates which reactions occur and gives a measure of the reaction rates as a function of catalyst temperature.

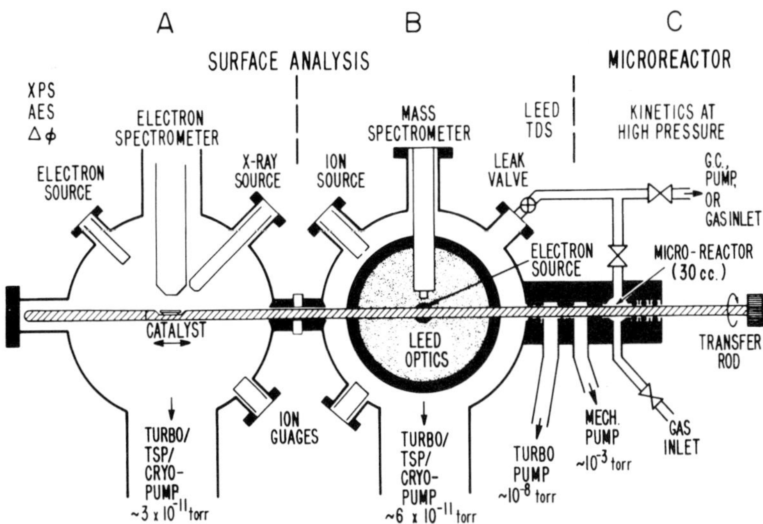

FIGURE 15. Schematic of apparatus combining UHV surface analytical chambers (Stages A and B) with a microreactor (C). (From Campbell, C. T. and Paffett, M. T., *Surf. Sci.*, 139, 396, 1984. With permission.)

This has now been effectively used for examining the kinetics and mechanisms of surface reactions on single crystals. Similarly, TPRx data can be analyzed to provide qualitative (characteristic peak temperature of a state and the distribution of states) and quantitative kinetic information. Recently, Falconer and Schwartz[119] have prepared an excellent review of applications of TPD, TPSR, and TPRx to catalyst research.

I. Current Trends

As presented in the preceding section, the commonly used outstanding single-crystal surface diagnostic techniques (LEED, HREELS, AES, XPS, SIMS, and ISS) function only under UHV conditions, which are used to investigate the structure, chemisorption, and chemical reaction. However, the same catalytic reactions generally occur with observable rates only under high pressures (1 to 100 atm); the pressure dependence of the reaction kinetics is then determined (see subsequent section).

The surface structure and the catalytic performance that are determined on the single-crystal surfaces are compared for those measured for practical catalyst systems. The major obstacles in the application of this approach have been in bridging the pressure gap. However, the recent development of transfer techniques to move the sample back and forth between UHV surface analysis and high-pressure reaction conditions has, to some extent, alleviated this pressure-gap problem.[120-125]

Figure 15 illustrates one such assembly (transfer technology) developed by Campbell's group.[124] This apparatus involved UHV stages for surface preparation and pre- and post-reaction surface analyses, together with a catalytic microreactor for high pressure (up to 10 atm). The sample is readily translated between these stages to minimize surface chemical changes between preparation, reaction, and surface analysis. Together, these techniques can clarify (1) the elemental composition and oxidation states at the surface, (2) the types of bonding at the surface, (3) two-dimensional surface crystallography, (4) the nature and quantity of adsorbed molecules, and (5) the mechanisms, energetics, and kinetics of adsorption/desorption processes. With this combined synergetic approach, the kinetics of reactions catalyzed over well-defined surfaces at high and low pressures can be correlated with the adsorption characteristics of the reactants as well as with the kinetics of the same reaction catalyzed over practical catalysts.

II. KINETICS: METHODOLOGY

Generally, the kinetics of catalytic reactions are determined by measuring the concentrations of reactants and products as a function of time, temperature, and pressure or as a function of other reaction variables. Detection is usually made by means of a liquid or gas chromatograph or mass spectrometer and the reactions may be carried out in a flow or batch mode. Reaction temperature in conjuction with feed ratio are the most reliable variables in controlling the reaction system.

Rate data may be obtained in either batch or flow reactor. In the former case, the reactants are charged in bulk to a stirred vessel and measurements are made of the course of reaction, whereas in the latter type, reactants are charged continuously at calculated rates through a comparatively long, narrow tube and observations are made when a steady state is reached. The tubular flow reactor may be a differential type, which is so short that only a small, though necessarily measurable, amount of conversion takes place, thus affording direct evaluation of the rate; or it may be integral, in which a comparably large conversion may take place, thus providing conversion vs. contact time.

Before the bulk of the experimental work is undertaken, certain preliminary studies are made to aid in an analysis of the results. The experimental conditions are so chosen that the reaction rate is not influenced by external and internal diffusion. To establish this, the influence of space velocity, the ratio of the weight of the catalyst (W) to the molar feed flow rate (F), catalyst mass, and particle size are studied. The effects of diffusion are checked by passing the gas at high velocity through the catalyst. The fair constancy of conversion is obtained by changing the feed rate while keeping (W/F) constant which suggests that the diffusion of the gases are not controlling the rate. The role of internal diffusion is checked by varying the size of the catalyst particles while maintaining the catalyst weight and (W/F) constant. The insignificant change in the reaction rate suggests that the internal diffusion resistance is negligible. To study the effect of external diffusion, it is necessary to increase the bed height while keeping (W/F) constant. In this way, the influence of external diffusion in the catalyst bed is determined. Except for direct test, there appears to be no positive way to establish the importance of diffusion in any specific case.

For kinetic analysis, the effect of various variables, reactant mole ratio, reaction temperature, and (W/F), on the conversion are investigated. An equation for the overall rate of reaction embodying the rates of the steps of surface-reaction kinetic models (chemisorption, desorption, or surface chemical reaction) would be quite formidable, and the rate constants would be so inextricably bound up with the variables that they would be difficult to evaluate accurately. Usually, it is hoped that only one of the steps offers appreciable resistance to the reaction.

The experimental rate data are analyzed with the use of regression techniques for model discrimination and parameter estimation. Nonlinear regressions[126-129] are generally carried out in order to obtain a mathematical fit for various rate expressions derived from LHHW or redox mechanisms. The derivation of models and details of model discrimination and parameter estimation procedures have already been described in the preceding chapter.

For determining the kinetics of deactivation, the deactivation rate data are taken from the experiments carried out in a fixed-bed reactor for several hours at various operating conditions (feed ratios, W/F, and temperatures).[84] A plot of conversion (operation time) is obtained. From the conversion vs. W/F curves at different operation times, the reaction rate, which is the slope of the curve, is calculated for a particular space time (W/F), at different operation times. Then the activity at any time t is determined as the ratio between the reaction rate at any time t, $(-r)_j$, and the reaction rate at zero time, $(-r)_j o$, with both rates measured at the same conversion or partial pressure of the reaction mixture. The deactivation rates $(-da/dt)$ are evaluated by analytical differentiation. The deactivation function values are

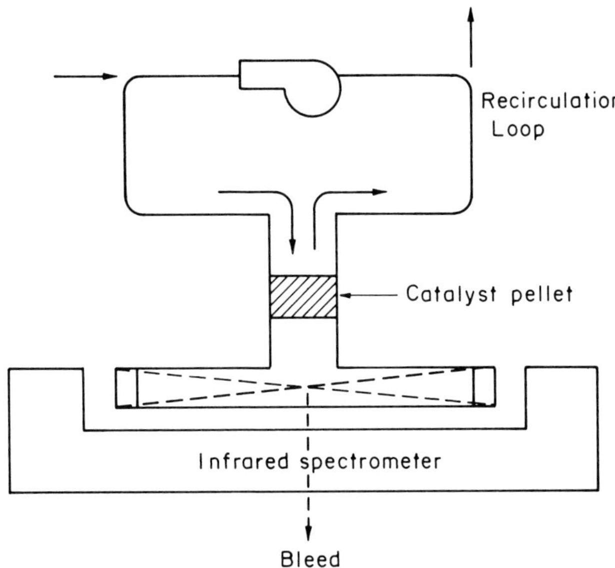

FIGURE 16. Schematic of Petersen's poisoned-pellet reactor. (From Hegedus, L. and Petersen, E. E., *Catal. Rev.*, 9, 245, 1974. Courtesy of Marcel Dekker, Inc.)

then calculated from Equation 9 (Chapter 1). These results are analyzed on the basis of LHHW kinetics with statistical data interpretation.

Deactivation kinetics have also been studied in a single-pellet reactor, developed by Petersen and co-workers,[130] which permits the direct assessment of catalytic reaction-deactivation processes. Reactants are found on one face of the pellet, while the other face is equivalent to the center plane of a pellet twice as long as the experimental pellet (Figure 16). By analyzing concentrations at the back face, centerline concentrations are secured, from which reaction-deactivation parameters can be extracted. However, the analysis with the use of a single-pellet diffusion reactor requires an accurate measurement of center-plane concentration.

REFERENCES

1. **Hair, M. L.,** *Infrared Spectroscopy in Surface Chemistry,* Marcel Dekker, New York, 1967.
2. **Blyholder, G.,** in *Experimental Methods in Catalytic Research,* Anderson, R. B., Ed., Academic Press, New York, 1968.
3. **Force, E. L. and Bell, A. T.,** *J. Catal.,* 40, 356, 1975.
4. **Eischens, R. P. and Pliskin, W. A.,** *Adv. Catal.,* 10, 1, 1958.
5. **Dumesic, J. A. and Topsoe, H.,** *Adv. Catal.,* 26, 122, 1977.
6. **Lunsford, J. H.,** *Adv. Catal.,* 22, 265, 1972.
7. **Srivastava, R. D., Stiles, A. B., and Jones, G. A.,** *J. Catal.,* 77, 192, 1982.
8. **Schultz, J. M.,** *Diffraction for Material Scientists,* Prentice-Hall, Englewood Cliffs, N.J., 1982.
9. **Werner, H. W. and Garten, R. P. H.,** *Rep. Prog. Phys.,* 47, 221, 1984.
10. **Spencer, N. D. and Somorjai, G. A.,** *Rep. Prog. Phys.,* 46, 1, 1983.
11. **Koestner, R. J., Van Hove, M. A., and Somorjai, G. A.,** *J. Phys. Chem.,* 87, 203, 1983.
12. **Marcus, P. M. and Jona, F.,** *Appl. Surf. Sci.,* 11/12, 20, 1982.
13. **Poate, J. M. and Buck, T. M.,** *Experimental Methods in Catalytic Research,* Vol. 3, Anderson, R. B. and Dawson, P. T., Eds., Academic Press, New York, 1976, chap. 5.

14. **Baun, W. L.,** *Surf. Interface Anal.*, 3, 243, 1981.
15. **Buck, T. M.,** *Inelastic Ion-Surface Collisions,* Tolk, N. H., Tully, J. C., Heiland, W., and White, C. W., Eds., Academic Press, New York, 1977, 47.
16. **Benninghoven, A.,** *Secondary Ion Mass Spectrometry, SIMS III,* Springer-Verlag, Berlin, 1982, 438.
17. **Park, R. L.,** *Experimental Methods in Catalytic Research,* Vol. 3, Anderson, R. B. and Dawson, P. T., Eds., Academic Press, New York, 1976, 1.
18. **Briggs, D.,** *Electron Spectroscopy: Theory, Techniques and Applications,* Vol. 3, Brundle, C. R. and Baker, A. D., Eds., Academic Press, New York, 1979, 305.
19. **Ponec, V.,** *Mater. Sci. Eng.*, 42, 135, 1980.
20. **Thomas, J. M. and Lambert, R. M.,** *Characterization of Catalysts,* John Wiley & Sons, Chichester, U.K., 1980.
21. **Haller, G. T.,** *Catal. Rev. Sci. Eng.*, 23, 477, 1981.
22. **McGuire, G. E. and Holloway, P. H.,** *Electron Spectroscopy; Theory, Techniques and Applications,* Vol 4, Brundle, C. R. and Baker, A. D., Eds., Academic Press, New York, 1981, 274.
23. **Bhasin, M. M.,** *Chem. Eng. Prog.*, 77(3), 60, 1981.
24. **Zhdan, P. A.,** *Stud. Surf. Sci. Catal.*, 9, 191, 1982.
25. **Hercules, D. M. and Klein, J. C.,** *Anal. Chem.*, 52, 1803, 1982.
26. **Bonzel, H. P.,** *Fresenius Z. Anal. Chem.*, 314, 310, 1983.
27. **Delannay, F.,** *Characterization of Hetergeneous Catalysts,* Marcel Dekker, New York, 1984.
28. **Eland, J. H. D.,** *Photoelectron and Auger Spectroscopy,* Butterworths, London, 1974.
29. **Carlson, T. A.,** *Photoelectron and Auger Spectroscopy,* Plenum Press, New York, 1975.
30. **Carlson, T. A.,** *Surf. Interface Anal.*, 4, 125, 1982.
31. **Baker, A. D. and Betteridge, D.,** *Photoelectron Spectroscopy,* Pergamon Press, New York, 1972.
32. **Brundle, C. R. and Baker, A. D.,** *Electron Spectroscopy; Theory, Techniques and Applications,* Academic Press, New York, 1977.
33. **Siegbahn, K., Nordling, C., Fahlmann, A., Nordberg, R., Hamrin, K., Hedman, J., Johansson, G., Bergmark, T., Karlsson, S. E., Lindgern, I., and Lindberg, B.,** *ESCA-Atomic, Molecular and Solid State Structure Studied by means of Electron Spectroscopy,* Nova Acta Regiae Societatis Scientarium Upsaliensis Ser. IV, Vol 20, Uppsala: Almqvist and Wiksells Boktryckeri AB, 1967.
34. **Hercules, S. H. and Hercules, D. M.,** *Characterization of Solid Surfaces,* Kane, P. F. and Larrabee, G. B., Eds., Plenum Press, New York, 1974, 307.
35. **Holm, R. and Storp, S.,** Analysen und Messverfahren, *Ullmanns Encyclopadie der Technischen Chemie, Band 5,* Kelkar, H., Ed., Verlag Chemie, Weinheim, 1980, 519.
36. **Jolly, W. L.,** *Electron Spectroscopy: Theory, Techniques and Applications,* Vol. 1, Brundle, C. R. and Baker, A. D., Eds., Academic Press, New York, 1977, 119.
37. **Fadley, C. S.,** *Electron Spectroscopy: Theory, Techniques and Applications,* Vol. 2, Brundle, C. R. and Baker, A. D., Eds., Academic Press, New York, 1978, 2.
38. **Baltanas, M. A.,** Ph.D. dissertation, University of Delaware, Newark, 1982.
39. **Srivastava, R. D., Onuferko, J., Schultz, J. M., Jones, G. A., Rai, K. N., and Athappan, R.,** *I & E C Fundam.*, 21, 457, 1982.
40. **Seah, M. P.,** *Surf. Interface Anal.*, 1, 86, 1979.
41. **Davis, L. E., MacDonald, N. C., Palmberg, P. W., Riach, G. E., and Weber, R. E.,** *Handbook of Auger Electron Spectroscopy,* Perkin-Elmer, Eden Prairie, Minn., 1978.
42. **Chang, C. C.,** *Characterization of Solid Surfaces,* Kane, P. F. and Larrabee, G. B., Eds., Plenum Press, New York, 1974, 509.
43. **Grant, J. T.,** *Appl. Surf. Sci.*, 13, 35, 1982.
44. **Hofmann, S.,** *Wilson and Wilson's Comprehensive Analytical Chemistry,* Vol. 9, Svehla, G., Ed., Elsevier, Amsterdam, 1979, 89.
45. **Joshi, A., Davis, L. E., and Palmberg, P. W.,** *Methods of Surface Analysis,* Czanderna, A. W., Ed., Elsevier, Amsterdam 1975, 159.
46. **Sinfelt, J. H., Via, G. H., and Lytle, F. W.,** *J. Chem. Phys.*, 72, 4832, 1980.
47. **Sinfelt, J. H., Via, G. H., Lytle, F. W., and Greegor, R. B.,** *J. Chem. Phys.*, 75, 5527, 1981.
48. **Sinfelt, J. H., Via, G. H., and Lytle, F. W.,** *J. Chem. Phys.*, 76, 2779, 1982.
49. **Lornston, J. M.,** Ph.D. dissertation, University of Delaware, Newark, 1980.
50. **Short, D. R., Mansour, A. N., Cook, J. W., Sayers, D. E., and Katzer, J. R.,** *J. Catal.*, 82, 299, 1983.
51. **Sand, M.,** Ph.D. dissertation, University of Delaware, Newark, 1982.
52. **Winick, H. and Bienestock, A.,** *Annu. Rev. Nucl. Part. Sci.*, 28, 33, 1978.
53. **Sandstorm, D. R. and Lytle, F. W.,** *Annu. Rev. Phys. Chem.*, 30, 215, 1979.
54. **Stern, E. A.,** *Contemp. Phys.*, 19, 289, 1978.
55. **Lytle, F. W., Via, G. H., and Sinfelt, J. H.,** *Synchroton Radiation Research,* Winick, H. and Doniach, S., Eds., Plenum Press, New York, 1980, 401.

56. **Eisenberger, P. and Kincaid, B. M.,** *Science,* 200(4389), 1441, 1978.
57. **Sayers, D. E., Lytle, F. W., and Stern, E. A.,** *Advances in X-Ray Analysis,* Vol. 13, Henke, B. L., Newkirk, J. B., and Mallett, G. R., Eds., Plenum Press, New York, 1970, 248.
58. **Sayers, D. E., Stern, E. A., and Lytle, F. W.,** *Phys. Rev. Lett.,* 27, 1204, 1971.
59. **Lytle, F. W., Sayers, D. E., and Stern, E. A.,** *Phys. Rev. B,* 11, 4836, 1975.
60. **Stern, E.,** *Phys. Rev. B,* 10, 3027, 1974.
61. **Ashley, C. A. and Doniach, S.,** *Phys. Rev. B,* 11, 1279, 1975.
62. **Lee, P. A. and Pendree, J. B.,** *Phys. Rev. B,* 11, 27, 1975.
63. **Lee, P. A. and Beni, G.,** *Phys. Rev. B,* 15, 2862, 1977.
64. **Via, G. H., Sinfelt, J. H., and Lytle, F. W.,** *J. Chem. Phys.,* 71, 690, 1979.
65. **Teo, B. K. and Lee, P. A.,** *J. Am. Chem. Soc.,* 101(11), 2815, 1979.
66. **Sinfelt, J. H.,** *Bimetallic Catalysts,* John Wiley & Sons, New York, 1983.
67. **Cairns, J. A.,** *Characterization of Catalysts,* Thomas, J. M. and Lambert, R. M., Eds., John Wiley & Sons, Chichester, U.K., 1980, chap. 2.
68. **Cairns, J. A., Lurio, A., Ziegler, J. F., Holloway, D. F., and Cookson, J. A.,** *J. Catal.,* 45, 6, 1976.
69. **Jothimurugesan, K., Nayak, A. K., Mehta, G. K., Rai, K. N., Bhatia, S., and Srivastava, R. D.,** *AIChE J.,* 31, 1997, 1985.
70. **Rosenfarb, J., Laitinen, H. A., Sanders, J. T., and Van Rinsvelt, H. A.,** *Anal. Chim. Acta,* 108, 119, 1979.
71. **Johansson, T. B., Akselsson, R., and Johansson, S. A. E.,** *Nucl. Instr. Meth.,* 84, 141, 1970.
72. **Ahlberg, M., Johanson, G., and Malmquist, K.,** *Nucl. Instr. Meth.,* 131, 371, 1975.
73. **Ziegler, J. F.,** *New Uses of Ion Acclerators,* Plenum Press, New York, 1975.
74. **Chu, W. K., Mayer, J. W., and Nicolet, M. A.,** *Backscattering Spectrometry,* Academic Press, New York, 1978.
75. **Buch, J. and Balalikin, N. L.,** *Mikrochim. Acta,* 1, 19, 1978.
76. **Cairns, J. A., Baglin, J. E. E., Clark, G. J., and Ziegler, J. F.,** *J. Catal.,* 83, 301, 1983.
77. **Srivastava, R. D., Prasad, N. S., and Pal, A. K.,** *Chem. Eng. Sci.,* 41, 719, 1986.
78. **Jothimurugesan, K.,** Ph.D. dissertation, Indian Institute of Technology, Kanpur, 1985.
79. **Smith, A. L.,** *Applied Infrared Spectroscopy,* John Wiley & Sons, New York, 1979.
80. **Barth, R., Gates, B. C., Zhao, Y., and Knozinger, H.,** *J. Catal.,* 82, 147, 1983.
81. **Alpert, N. L., Keiser, W. E., and Szymanski, H. A.,** *Theory and Practice of Infrared Spectroscopy,* Plenum Press, New York, 1970.
82. **Long, D. A.,** *Raman Spectroscopy,* McGraw-Hill, New York, 1977.
83a. **Griffiths, P. R.,** *Chemical Fourier Transform Spectroscopy,* John Wiley & Sons, New York, 1975.
83b. **Griffiths, P. R.,** in *Advances in Infrared Raman Spectroscopy,* Vol. 10, Clark, R. J. and Hester, R. E., Eds., Heyden, London, 1983, 277.
84. **Backx, C., De Groot, C. P. M., and Biloen, P.,** *Appl. Surf. Sci.,* 6, 256, 1980.
85. **Frauenfelder, H.,** *The Mössbauer Effect,* W. A. Benjamin, New York, 1962.
86. **Wertheim, G. K.,** *Mössbauer Effect; Principles and Applications,* Academic Press, New York, 1964.
87. **Gager, H. M. and Hobson, M. C.,** *Catal. Rev.,* 11, 117, 1975.
88. **Brongersma, H. H.,** *J. Vac. Sci. Technol.,* 11, 231, 1974.
89. **Goff, R. F. and Smith, D. P.,** *J. Vac. Sci. Technol.,* 7, 72, 1970.
90. **Taglauer, E. and Heiland, W.,** *Appl. Phys.,* 9, 261, 1976.
91. **Kelley, M. J. and Ponec, V.,** *Prog. Surf. Sci.,* 11, 139, 1981.
92. **Kelley, M. J., Freed, R. L., and Swartzfager, D. G.,** *J. Catal.,* 78, 445, 1982.
93. **Kelley, M. J., Short, D. R., and Swartzfager, D. G.,** *J. Mol. Catal.,* 20, 235, 1983.
94. **Campbell, C. T.,** Growth modes of Cu on Ag(111) studied by ISS, to be published.
95. **Buhl, R. and Preisinger, A.,** *Surf. Sci.,* 47, 344, 1975.
96. **Suib, S. L., Coughlin, D. F., Otter, F. A., and Conopask, L. F.,** *J. Catal.,* 84, 410, 1983.
97. **Chin, R. L. and Hercules, D. M.,** *J. Phys. Chem.,* 86, 360, 1982.
98. **Pierce, J. L. and Walton, R. L.,** *J. Catal.,* 81, 375, 1983.
99. **Marien, J., DePauw, E., and Peizer, G.,** *J. Phys. Chem.,* 87, 4344, 1983.
100. **Creighton, J. R. and White, J. M.,** *Surf. Sci.,* 129, 327, 1983.
101. **Drechsler, M., Hoinkes, H., Kaarmann, H., Wilsch, H., Ertl, G., and Weiss, M.,** *Appl. Surf. Sci.,* 3, 217, 1979.
102. **Barteau, M. A. and Madix, R. J.,** in *The Chemical Physics of Solid Surfaces and Heterogeneous Catalysis,* King, D. A. and Woodruff, D. P., Eds., Elsevier, Amsterdam, 1982, chap. 4.
103. **McRae, E. G.,** *Rev. Mod. Phys.,* 51, 541, 1979.
104. **Thomas, G. E. and Weinberg, W. H.,** *J. Chem. Phys.,* 70, 1000, 1979.
105. **Persson, B. N. J.,** *Solid State Commun.,* 24, 573, 1977.
106. **Andersson, S. and Davenport, J. W.,** *Solid State Commun.,* 28, 677, 1978.

107. **Ibach, H.,** *Surf. Sci.*, 66, 56, 1977.
108. **Fisher, G. B. and Sexton, B. A.,** *Phys. Rev. Lett.*, 44, 683, 1980.
109. **Lehwald, S., Erley, W., Ibach, H., and Wagner, H.,** *Chem. Phys. Lett.*, 62, 360, 1979.
110. **Barnes, M. R. and Willis, R. F.,** *Phys. Rev. Lett.*, 41, 1729, 1978.
111. **Backx, C. and Willis, R. F.,** *Chem. Phys. Lett.*, 53, 471, 1978.
112. **Sexton, B. A.,** *Surf. Sci.*, 88, 319, 1979.
113. **Demuth, J. E. and Ibach, H.,** *Surf. Sci.*, 85, 365, 1979.
114. **Somorjai, G. A.,** *Acc. Chem. Res.*, 9, 248, 1976.
115. **King, D. A.,** *CRC Crit. Rev. Solid State Mater. Sci.*, 7(3), 1978.
116. **Barteau, M. A., and Madix, R. J.,** *Surf. Sci.*, 97, 101, 1980.
117. **Kiskinova, M., Griffin, G. L., and Yates, J. T.,** *J. Catal.*, 71, 278, 1981.
118. **Cvetanovic, R. J. and Amenomiya, Y.,** in *Advances in Catalysis and Related Subjects*, Vol. 17, Academic Press, New York, 1967, 103.
119. **Falconer, J. L. and Schwartz, J. A.,** *Catal. Rev. Sci. Eng.*, 25, 141, 1983.
120. **Gillespie, W. D., Herz, R. K., Petersen, E. E., and Somorjai, G. A.,** *J. Catal.*, 70, 147, 1981.
121. **Sexton, B. A. and Somorjai, G. A.,** *J. Catal.*, 46, 167, 1977.
122. **Goodman, D. W., Kelley, R. D., Madey, T. E., and Yates, J. T.,** *J. Catal.*, 63, 226, 1980.
123. **Bonzel, H. P. and Krebs, H. J.,** *Surf. Sci.*, 88, 269, 1979.
124. **Campbell, C. T. and Paffett, M. T.,** *Surf. Sci.*, 139, 396, 1984.
125. **Davis, S. M. and Somorjai, G. A.,** in *The Chemical Physics of Solid Surfaces and Heterogeneous Catalysis*, King, D. A. and Woodruff, D. P., Eds., Elsevier, Amsterdam, 1982, chap. 7.
126. **Draper, N. R. and Smith, H.,** *Applied Regression Analysis*, John Wiley & Sons, New York, 1966.
127. **Athappan, R. and Srivastava, R. D.,** *AIChE J.*, 26, 517, 1980.
128. **Goyal, H. B., Kudchadkar, A. P., and Srivastava, R. D.,** *J. Appl. Chem. Biotechnol.*, 28, 547, 1978.
129. **Guha, A. K. and Srivastava, R. D.,** *J. Catal.*, 91, 254, 1985.
130. **Hegedus, L. and Petersen, E. E.,** *Catal. Rev.*, 9, 245, 1974.
131. **Topsoe, H., Clausen, B. S., Candia, R., Wivel, C., and Morup, S.,** *J. Catal.*, 68, 433, 1981.
132. **Wivel, C., Candia, R., Clausen, B. S., Morup, S., and Topsoe, H.,** *J. Catal.*, 68, 453, 1981.

Chapter 3

SELECTIVE OXIDATION OF HYDROCARBONS

I. GENERAL

The selective or partial oxidation is a commercial process for converting relatively cheap feed stocks as hydrocarbons into valuable chemical products for the petrochemical industry. The term 'selective' indicates that the oxidation reaction leads to one special product; its rate of production being favored over others, in particular, the total oxidation to CO or CO_2. The oxidation can be carried out in the gas phase over heterogeneous catalysts (transition-metal oxides) or in the liquid phase with the use of homogeneous catalysts (coordination complexes of transition metals).

In recent years it has been found that homogeneous catalysts consisting of transition-metal complexes also catalyze a variety of reactions with high selectivity and activity. The commercial importance of such catalysts has been firmly established by the success of the Wacker® process for acetaldehyde or vinyl acetate[1-3] and of the oxo process for hydroformulation of α-olefins[4] as well as the Monsanto® process for acetic acid.[5] Heterogeneous oxidation, however, has the advantage of simple separation of catalysts and products while for homogeneous oxidation processes this always presents a problem.

A solid catalyst for a partial oxidation process is designed to provide a limited amount of active oxygen to a reactant, allowing formation of the desired product, but restricting further oxidation that would give rise to carbon monoxide or carbon dioxide, which is now a major tool for the incorporation of groups such as carbonyl, aldehyde, and nitrile in hydrocarbons. Catalytic oxidation is necessary to obtain selective products, since carbon oxides are the thermodynamically favored products. For example, the complete oxidation of propene and ammonia to carbon dioxide, nitrogen, and water at temperatures of interest is preferred to the formation of acrylonitrile or acrolein.

In general, the catalyst selectivity is often sensitive to reaction conditions, especially temperature. Since the reactions are strongly exothermic, the reactor temperature tends to increase with increasing distance downstream of the inlet. But, as the reactants proceed through the reactor they are depleted, so that the rate of reaction (and heat evolution) may decrease toward the exit; the temperature profile may show a maximum, the location of which is referred to as a hot spot. As a result, the reactor temperature profiles are difficult to predict and the reactor design, therefore, becomes difficult. Thus, an important factor which is now considered in the optimization procedures is the effect of small perturbations in the operating parameters, such as feed temperature, coolant temperature, and feed composition. It has been shown that such small changes can severely affect the operation of the reactor by causing a large increase in the hot-spot temperature. Under such conditions, the reactor is considered to be 'parametrically sensitive'.

For the foregoing heterogeneous catalytic oxidation reactions, fluidized-bed reactors are commonly employed. A fluidized-bed reactor is one in which relatively small particles of catalyst (10 to 300 μm) are suspended by the upward motion of the reacting gas. The gas enters the bottom of the column through a support plate or distributor and flows through the bed to exit from the top of the column. The particles are in constant motion within a relatively confined region of space, and extensive mixing occurs in both the radial and longitudinal directions of the bed. The circulation of the bed and uniform agitation within it prevent the occurrence of hot spots and dead regions. It also makes possible the continuous circulation of the catalyst between the reaction vessel and a regeneration vessel. The major disadvantage of this system is that the catalyst is eroded and broken down by its constant

Table 1
INDUSTRIAL SELECTIVE OXIDATION PROCESSES

Reaction	Catalyst
Phthalic anhydride from naphthalene $C_{10}H_8 + 9/2O_2 = C_6H_4(CO)_2O + 2H_2O + 2CO_2$	Supported V_2O_5 with promoters
Phthalic anhydride from o-xylene $C_6H_4(CH_3)_2 + 3O_2 = C_6H_4(CO)_2O + 3H_2O$	Supported V_2O_5 with K_2SO_4
Maleic anhydride from benzene $C_6H_6 + 9/2O_2 = C_2H_2(CO)_2O + 2CO_2 + 2H_2O$	Supported V_2O_5 + MoO_3
Benzaldehyde from toluene $C_6H_5CH_3 + O_2 = C_6H_5CHO + H_2O$	Supported oxides of U and Mo
Formaldehyde from methanol $CH_3OH + 1/2O_2 = HCHO + H_2O$	Fe_2MoO_4 + MoO_3 or Ag gauge
Acrolein from propene $CH_2 = CH - CH_3 + O_2 = CH_2 = CH - CH = O + 3/2H_2O$	Bismuth molybdates
Acrylonitrile from ammoxidation of propene $CH_2 = CH - CH_3 + NH_3 + 3/2O_2 = CH_2 = CHCN + 3H_2O$	Multicomponent bismuth molybdates

motion, so that there is continual attrition of the particles requiring continual makeup of fresh catalyst. However, this can be minimized with cyclone separators, frequently mounted within the reactor itself.

An intensive research effort has been directed towards the development of selective catalysts for the partial oxidation of hydrocarbons to a number of products. This has led to the development of a number of selective oxidation catalysts made from transition-metal oxides. Many of the catalysts are not simple oxides but binary mixtures of compounds, and the binary systems have catalytic properties strikingly different from those of the components. For example, the selective oxidation of methanol to formaldehyde is carried out industrially over a catalyst consisting of a mixture of ferric molybdate, $Fe_2(MoO_4)_3$, and molybdenum trioxide, MoO_3. A typical question relates to which phase is the real catalyst. Neither phase by itself can be a practical catalyst. However, it is now known that iron tends to stabilize molybdenum oxide against volatilization and loss of surface area; the surface of ferric molybdate still gradually loses some molybdenum through volatilization, and then a loss in selectivity occurs. In the presence of excess molybdenum, this surface depletion does not occur. Nevertheless, there remains considerable controversy on the precise geometric relationship among the Mo atoms at the active sites. Some of the most important industrial heterogeneous catalytic oxidation processes employing supported metal oxides are listed in Table 1.

Most transition-metal oxide catalysts are characterized by complex and poorly defined surface structures, but better understanding of their surface structure, reaction paths, and mechanisms are now evolving. The applicability of the catalysts generally applies to an entire group of reactions, although the optimal system will probably be tailored for each specific reaction (Table 1).

Our lack of understanding limits the discussion to the two selective classes of oxidation catalysts for which a relatively coherent picture of catalyst structure and reaction mechanism has emerged; namely, (1) the oxidation reactions of hydrocarbons over V_2O_5-based catalysts, and (2) the oxidation and ammoxidation of propene over Bi-Mo-based catalysts to give acrolein and acrylonitrile, respectively. The later is considered to be a major event in the history of oxidation catalysis since this catalyst is very versatile, being effective in a number of processes, e.g., the oxidative dehydrogenation of *n*-butenes and methyl-butene to butadiene and isoprene, and the formation of aromatic carbonyls and nitriles. However, in order to discuss the catalytic oxidation on surfaces, it is useful to consider the chemistry of liquid-phase oxidations of hydrocarbons. A comprehensive review on the liquid-phase oxidations has recently appeared.[6]

In this chapter, V_2O_5- and Bi-Mo-catalyzed selective oxidations of hydrocarbons of the type practiced commercially will be considered, and some potential applications not currently in use will be introduced. These oxidation mechanisms may provide the bases for future industrial oxidation catalysis. The chapter proceeds with a brief consideration of the chemistry of liquid-phase oxidation of hydrocarbons.

II. LIQUID-PHASE OXIDATIONS

A common sequence for the oxidation of hydrocarbon involves the initial formation of hydroperoxide followed by its catalytic transformation to stable oxidation products *in situ*. The free radicals are initiated from a reaction between the catalyst (transition-metal salts or complexes) and the hydroperoxide involving a redox process with the metal. Initiation generally involves the breakdown of a catalyst-hydroperoxide complex, and termination of reaction chains by the catalyst that occur, to a variable extent, depending on the catalyst.

It is interesting to note that, for the heterogeneous liquid-phase oxidation of cyclohexene catalyzed by transition-metal oxides, the activity of the oxides was closely related to soluble catalysts of the same metal ion.[7] This led to the conclusion that the reaction mechanisms are similar for both types of catalysts.[8] Subsequently, Srivastava and co-workers[8-11] established that the catalytic activity of transition-metal oxides in the oxidations of cyclohexene, cumene, and tetralin is mainly due to their capabilities of decomposing hydroperoxide in the chain-initiating radicals. They concluded that such oxidations proceed via a degenerate chain-branching mechanism. The similarity in the kinetic behavior of such systems may be due to formation of resonance-stabilized species in each of these cases.

The only aromatics oxidized on a large scale industrially using cobalt catalysts are methyl aromatics. Reaction such as

$$\underset{CH_3}{\overset{CH_3}{\bigcirc}} \xrightarrow[\text{[Co]}]{O_2} \underset{COOH}{\overset{CH_3}{\bigcirc}} \xrightarrow[\text{[MEK,Co]}]{O_2} \underset{COOH}{\overset{COOH}{\bigcirc}} \qquad (1)$$

is an example of oxidation reaction catalyzed by transition-metal complex. As can be seen, the oxidation of *p*-xylene to terephthalic acid is limited by oxidation of the intermediate, *p*-toluic acid. This may be due to the ionization potential of *p*-toluic acid which is significantly higher than that of *p*-xylene, making oxidation via electron transfer more difficult; the ease of the electron-transfer oxidation of hydrocarbons to produce the cation radical is known to be related to the ionization potential of the organic substrate that is oxidized. For this reason, an additive such as methyl ethyl ketone (MEK) is used to promote complete oxidation to terephthalic acid by maintaining high concentrations of Co^{3+} species.[6,12]

This mechanism is only one example of a number of proposed mechanisms having several common characteristics, for example, oxidation of toluene to benzoic acid. In such cases, benzylic radical is the species that first interacts with molecular oxygen and subsequently forms stable oxidation products via a number of reaction sequences.

There are reasons to believe that free radicals are not always intermediates in oxidation reaction catalyzed by metal complexes. For example, in molybdenum-catalyzed oxidation of propene by tertiary butyl hydroperoxide, the hydroperoxide is activated by the molybdenum center and metal-oxygen-bonded species directly attack the olefin in a selective nonradical reaction:

$$CH_3CH{=}CH_2 + ROOH \xrightarrow{\text{catalyst}} CH_3{-}\overset{\displaystyle O}{CH{-}CH_2} + ROH \qquad (2)$$

$$R{=}(CH_3)_3C{-}$$

This reaction of propene with an organic hydroperoxide in the presence of metal complex forms the basis of commercial process for the production of propene oxide. In this class, an active catalyst contains the metal in its highest oxidation state, e.g., Mo(VI), W(VI), Ti(IV), and V(V).[6]

A transition-metal-complex-catalyzed reaction, the Wacker® reaction, uses Pd(II) to activate ethene toward nucleophilic attack by water and requires copper(II) to return Pd to the catalytically active state after β-hydride elimination. The reaction proceeds via the sequence:

$$PdCl_4^{2-} + C_2H_4 + H_2O \rightarrow CH_3CHO + Pd^0 + 2HCl + 2Cl^- \qquad (3)$$

$$Pd^0 + 2CuCl_2 + 2Cl^- \rightarrow PdCl_4^{2-} + 2CuCl \qquad (4)$$

$$2CuCl + 1/2O_2 + 2HCl \rightarrow 2CuCl_2 + H_2O \qquad (5)$$

The commercial process uses palladium chloride in a homogeneous liquid-phase reaction and may proceed either in a one- or two-stage process. In the former case, the olefin oxidation and catalyst reoxidation take place in the same reactor whereas in the latter, ethene is first oxidized and the acetaldehyde separated before cuprous chloride is reacted with oxygen.[13] Oxidants other than $CuCl_2$ can be used to regenerate Pd(II) in systems containing chloride ion.[6]

To illustrate some common features of catalytic oxidation in solutions and on surfaces, it is useful to consider the Pd-doped V_2O_5-Al_2O_3 heterogeneously catalyzed vapor-phase oxidation of ethene to acetaldehyde, analog to the Wacker® reaction, of Evnin et al.[14] Results from ESR spectra established a redox mechanism with the palladium ions playing a central role. The mechanism shown in Figure 1 was suggested to account coordination of both an olefin and a hydroxyl to Pd(II) followed by direct formation of the carbonyl compound and the reduction of the V_2O_5 substrate via the palladium center and finally reoxidation of the vanadium oxide by oxygen. It has also been reported by Seoane et al.[15] that Pd added to V_2O_5 only enhanced the reducibility of this oxide. From Figure 1, it appears that more electrons are needed to convert the O_2 molecule completely, which must be supplied by semiconduction bulk of the catalyst.

Based on these observations, the following conclusions, similar to Gates et al.,[16] are summarized because of their importance to the subsequent discussion of surface-catalyzed oxidation:

1. Many metal complexes catalyze the homolytic decomposition of hydroperoxides and initiate the oxidation, the specific nature of the metal compound used can determine the type of products that predominate. On the other hand, the transition-metal complex may directly interact with the hydrocarbon to effect a unique and selective oxidation.
2. It is sometimes advantageous to apply two different metal ions, one to attack the molecule to be oxidized and the other to interact with oxygen and facilitate electron transfer between the cations.
3. The oxygen introduced into the molecule which is oxidized need not necessarily be derived directly from O_2.

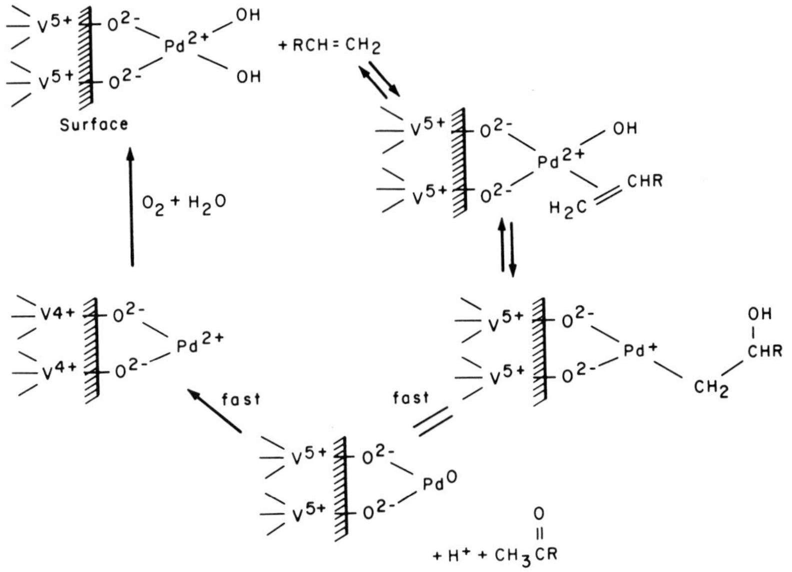

FIGURE 1. Mechanism of the Wacker reaction catalyzed by Pd-doped V_2O_5. (From Evnin, A. B., Rabo, J. A., and Kasai, P. H., *J. Catal.*, 30, 109, 1973. With permission.)

III. VANADIUM PENTOXIDE-BASED CATALYSTS

Supported V_2O_5 (10%) has long been one of the best oxidation catalysts for a number of industrial partial-oxidation processes like those of naphthalene, xylene, and benzene (Table 1). In the 1970s, due to economic advantage, the production of maleic anhydride from butane over vanadium-phosphorus-oxide catalysts (VPO) received much attention, and is beginning to replace the more wasteful production of maleic anhydride from benzene which is still the major feed stock. Research has advanced significantly, however, a great deal of fundamental work is required to analyze the role of VPO catalysts in the complex reaction network of selective transformation of *n*-butane to maleic anhydride. The subject has been reviewed recently by Hodnett.[56]

V_2O_5 has a zig-zag chain structure which provides inherent structural isolation. The addition of alkali metal sulfates and oxides enhances the selectivity by modification of solid-state structure. Their promoting action is found to increase from Li and Cs. For example, for K_2O, the conversion to selective product on K_2O/V_2O_5 ratio gives a maximum at K/V ≈ 0.25.

An electron micrograph of a typical K_2SO_4-promoted V_2O_5-SiO_2 is shown in Figure 2. In addition to small particles of V_2O_5 and K_2SO_4 on silica, some needle-type crystals are also observed. Except small particles and these needle-type structures, no other form of V_2O_5 was detected.[17,18] The electron diffraction pattern was found to be indexible on the basis of orthorhombic V_2O_5 and hexagonal K_2SO_4.

In the following pages the reaction kinetics of oxidation of hydrocarbons and alcohols are discussed. The discussion focuses on a generalized mechanism for such reactions.

A. Reaction Kinetics

The kinetics of V_2O_5-catalyzed oxidation of hydrocarbons and alcohols have been widely studied, and there is good agreement on this subject. From results of kinetics experiments on the oxidation of naphthalene, Mars and Van Krevelen[19] first proposed a two-stage redox mechanism. The reaction was supposed to occur in two steps: in the first step, the hydrocarbon

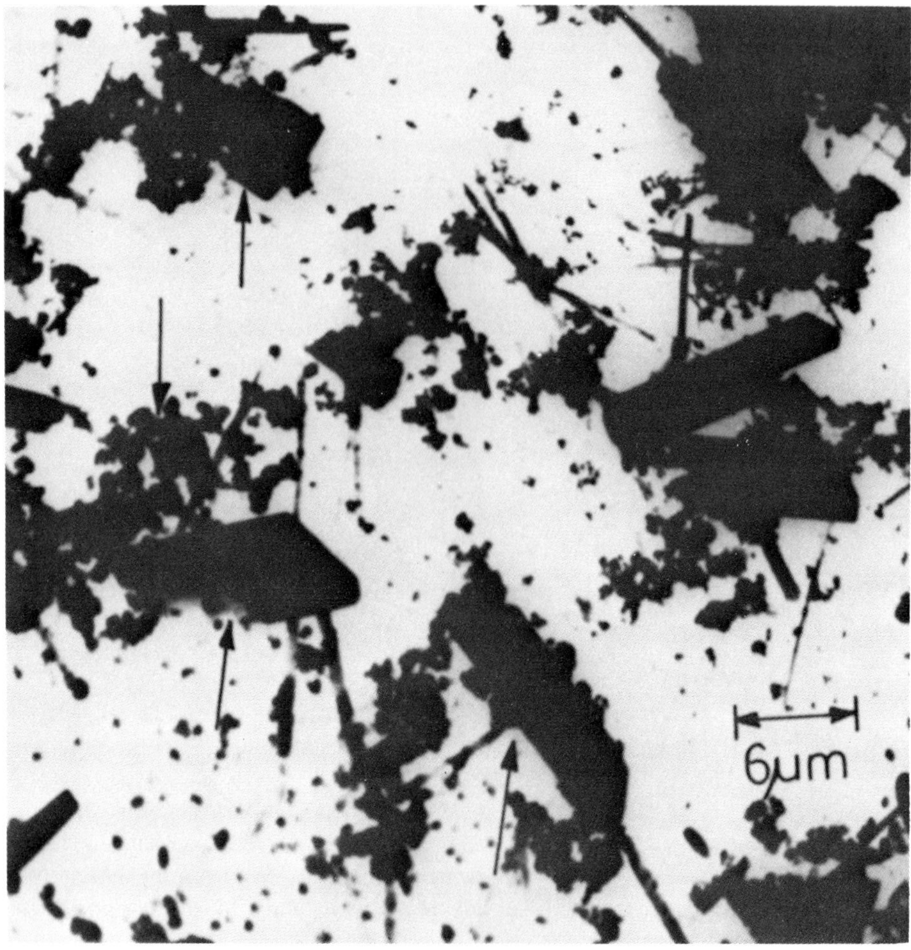

FIGURE 2. Electron micrograph showing V_2O_5 needles. (From Sharma, R. K., Rai, K. N., and Srivastava, R. D., *J. Catal.*, 63, 271, 1980. With permission.)

is oxidized and the oxide is reduced, while in the second step, there is a reaction between the reduced oxide and the oxygen in the gas phase to arrive at the initial state. The validity of this pattern was established for a series of oxidation reactions.

Sharma and Srivastava[20-22] studied in detail the kinetics of supported V_2O_5 catalyst by oxidizing toluene, methanol, and butanol to aldehydes and water, with the aim to propose a common theory for all such similar reactions. The experimental results were analyzed on the basis of Langmuir-Hinshelwood, modified Hinshelwood, and two-stage redox kinetics, with statistical data interpretation to show the real significance of mechanism determination with precise experimental data. This also serves as an excellent example of model discrimination and parameter estimation (Chapter 1). Their analysis is described below. The reactions under study were

$$\underset{(A)}{C_6H_5CH_3} + \underset{(B)}{O_2} \rightarrow \underset{(R)}{C_6H_5CHO} + \underset{(W)}{H_2O} \qquad (6)$$

$$CH_3OH + 1/2O_2 \rightarrow HCHO + H_2O \tag{7}$$

(A) (B) (R) (W)

$$C_4H_9OH + 1/2O_2 \rightarrow C_3H_7CHO + H_2O \tag{8}$$

(A) (B) (R) (W)

Reactions were carried out in a fixed bed-type reactor with a continuous flow system at atmospheric pressure and at three temperatures. Based on sequential experimentation design, a total of 45 experiments were performed for each isothermal set of runs. The experimental conditions were so chosen that aldehyde was the only product of oxidation. Introductory experiments were also carried out with mixed feed to study product inhibition to obtain enough information for discrimination between possible kinetic schemes and models. The data pointed out that oxygen pressure influenced the rates of all three reactions, while adsorption of water and aldehyde had no appreciable effect on the reaction rate.

The possible Langmuir-Hinshelwood isothermal rate equations based on single- and dual-site mechanisms for the three reactions were considered. These resulted in 46 different mathematical forms (for each system) which were confronted with the experimental data. A rate model derived on the basis of modified Hinshelwood mechanism was also confronted to the experimental data. Linear and nonlinear regressions of the rate equations were made at each temperature.

A two-stage redox mechanism was also adopted to include these oxidations. According to this mechanism, a steady state is assumed between the following two steps:

$$mA(g) + \text{Cat-O} \xrightarrow{k_1} R(g) + W(g) + \text{Cat}\cdot \tag{9}$$

$$nO_2(g) + \text{Cat}\cdot \xrightarrow{k_2} \text{Cat-O} \tag{10}$$

When the rates of both processes are equal,

$$r = \frac{k_1 p_A^m}{(1 + \alpha k_1 p_A^m / k_2 p_{O_2}^n)} \tag{11}$$

where m and n are the reaction order with respect to organic compound A and oxygen, respectively, and α is the number of oxygen molecules required to oxidize one molecule of organic compound to aldehyde. The isothermal models, for m and n between 0 and 1, were applied to the experimental data.

Table 2 gives the rate models with all positive constants which had to be retained statistically after isothermal regression. These models were common to all three reaction systems. The converged values of the parameters of these models by nonlinear estimation are presented in Table 3.

Upon applying the criteria established in the preceding chapters, it was evident that the oxidation process is not well described by any of the kinetic equations developed using Langmuir-Hinshelwood mechanism; SSAOD-4 model has the large confidence intervals for the various parameters and has been omitted because they are meaningless. However, modified Hinshelwood expression (model SSA-M) and redox model (VK-4) satisfactorily described the kinetics for the three systems.

Although both these mechanisms resulted in an identical rate expression, the significance

Table 2
MODELS REMAINING AFTER ISOTHERMAL REGRESSION

Model	Rate controlling step	Rate equation
	Langmuir-Hinshelwood	
SSAOD-4	Single-site surface reaction with A in the gas phase and oxygen dissociated (products not adsorbed)	$r = \frac{kK_Bp_Ap_B}{[1 + (K_Bp_B)^{1/2}]^2}$
	Modified Hinshelwood	
SSAM	Rate of adsorption of oxygen = rate of chemical reaction	$r = \frac{kk_Bp_Ap_B}{(k_Bp_B + \alpha kp_A)}$
	Redox Models	
VK-4	Reaction order: m(A) 1, n(B) 1	

Table 3
ISOTHERMAL REGRESSION

Model	Temp (°C)	Parameters	
SSAOD-4		k (gmol h^{-1} g^{-1} atm^{-1})	K_B (atm^{-1})
Toluene	320	(0.0799 ± 0.0073)	(125.9512 ± 97.6715)
	345	(0.0961 ± 0.0132)	(1487.2050 ± 2282.4467)
	370	(0.1153 ± 0.0145)	(972.1962 ± 632.1715)
Methanol	320	(0.0168 ± 0.0057)	(139.5301 ± 2775.0410)
	345	(0.1330 ± 0.0236)	(14.9860 ± 11.9080)
	370	(0.2414 ± 0.0377)	(8.7311 ± 2.3530)
Butanol	240	(0.0119 ± 0.0005)	(41.6332 ± 12.7803)
	260	(0.0121 ± 0.0008)	(48.9660 ± 22.4930)
	280	(0.0187 ± 0.0282)	(5.7934 ± 174.8015)
SSA-M		k (gmol h^{-1} g^{-1} atm^{-1})	k_B (gmol h^{-1} g^{-1} atm^{-1})
or		or	or
VK-4		k_1 (gmol h^{-1} g^{-1} atm^{-1})	k_2 (gmol h^{-1} g^{-1} atm^{-1})
Toluene	320	(0.0607 ± 0.0007)	(0.0011 ± 0.0041)
	345	(0.1430 ± 0.0098)	(0.0089 ± 0.0006)
	370	(0.5681 ± 0.0787)	(0.0153 ± 0.0012)
Methanol	320	(0.0159 ± 0.0056)	(0.0091 ± 0.0061)
	345	(0.0625 ± 0.0134)	(0.0571 ± 0.0321)
	370	(0.0897 ± 0.0153)	(0.0922 ± 0.0058)
Butanol	240	(0.0059 ± 0.0004)	(0.0022 ± 0.0006)
	260	(0.0060 ± 0.0002)	(0.0032 ± 0.0001)
	280	(0.0156 ± 0.0004)	(0.0034 ± 0.0002)

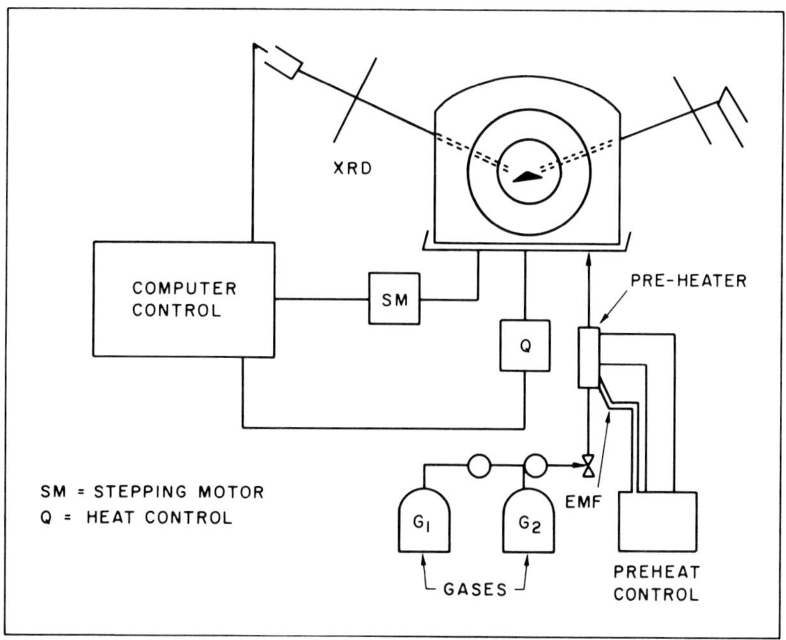

FIGURE 3. Schematic of *in situ* X-ray diffraction assembly.

of the parameters of the two models is different. The redox mechanism proposes the removal of actual lattice oxygen atoms by direct interaction with reductant, which must hence be chemisorbed, while Hinshelwood mechanism ignores such a catalyst-reductant interaction and instead, suggests that only adsorbed oxygen molecules or atoms (and not the lattice oxygen) are removed by reacting with the reductant molecules, which remain exclusively in the gas phase. Therefore, from a purely statistical point of view, it is impossible to say that one of these models described the data best. The discrimination between these two rival mechanisms must be based on the physicochemical insight and other experimental evidences. An inspiring verification of mechanism is offered by Srivastava et al.[23,24] in the *in situ* X-ray measurement of this catalyst-reaction system. Details of the structural dynamics of V_2O_5-SiO_2 catalysts follow.

B. Structural Dynamics

Investigations of the structural dynamics of V_2O_5-SiO_2 catalysts, when used for the selective vapor-phase oxidation reactions of organic compounds, have been recently attempted. A specially designed temperature-programmed *in situ* X-ray diffractometer has been used to follow the oxidation-state changes of a V_2O_5-SiO_2 catalyst in the oxidation of hydrocarbons.[23] This novel technique is of value in obtaining information about the exact phases present in the catalyst during the reaction.

The essential features of the experimental set-up are shown in Figure 3. *In situ* X-ray diffraction measurements of the catalyst are made at programmed 2θ and temperature intervals under different gas exposures. Figure 4 reveals the gradual reduction of V_2O_5 in toluene as the sample was heated from 25 to 300°C. A mixture of two phases — V_2O_5 and V_4O_9 — is evident at 200°C. No structural changes were reported in the SiO_2 catalyst support. Both exposures in Figure 5 represent steady-state conditions. As indicated, no reduction was obtained in air; however, expectedly, on carrying out the oxidation of toluene it was reported that V_4O_9 is predominant with a small amount of V_2O_5.[24] A similar mechanism prevailed in the silica-supported V_2O_5-catalyzed oxidation of other organic compounds.[23]

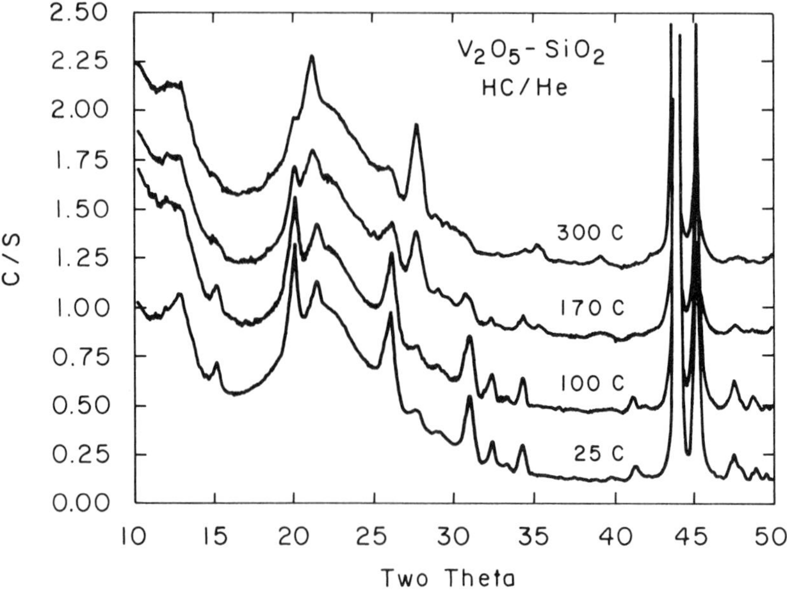

FIGURE 4. *In situ* X-ray diffraction patterns of V_2O_5-SiO_2 in toluene at various temperatures.

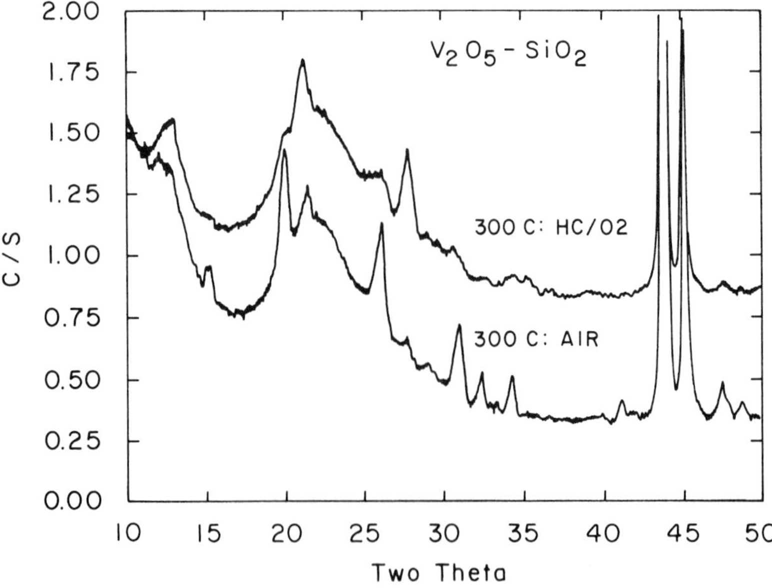

FIGURE 5. *In situ* X-ray diffraction patterns of V_2O_5-SiO_2 at steady state in oxidation of toluene.

XPS has been effectively employed to investigate the surface-oxidation states of transition metals. The V(2p) band of the vanadia-alumina catalyst before and after reaction with toluene is shown in Figure 6;[25] the V2(p) band of pure V_2O_5 is included for the purpose of comparison. Figure 6(b) showed the presence of a small proportion of V(IV) in addition to V(V), as characterized by the $V(2p_{3/2})$ signals at 515.6 and 516.5 eV, respectively. On carrying out the reaction with toluene, an increase in the proportion of V(IV) is revealed (Figure 6c). Such oxidation-state changes have also been described in the literature for a variety of

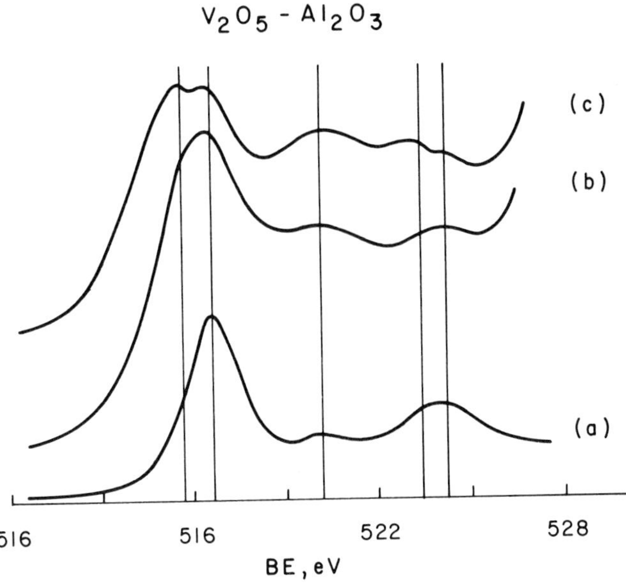

FIGURE 6. V (2p) bands in supported vanadia catalysts: (a) pure V_2O_5; (b) calcined V_2O_5-Al_2O_3; (c) V_2O_5-Al_2O_3 after reaction with toluene. (From Jagannathan, K., Srinivasan, A., and Rao, C. N. R., *J. Catal.*, 69, 418, 1981. With permission.)

reactions, for example, in recent papers concerning the oxidation of alkylpyridines on V_2O_5-TiO_2 and V_2O_5-SnO_2,[26,27] the ammoxidation of 3-picoline with V_2O_5-SnO_2,[28] and with other vanadium oxides.[15,29,30] We therefore conclude that the two-stage redox model is useful for this type of catalytic reaction.

IV. THE BI-MO-BASED CATALYST SYSTEM

As mentioned earlier in the text, a major event in the history of oxidation catalysis was the discovery of bismuth molybdate[31,32] as a selective catalyst for the partial oxidation of propene and also in one-step operation for the ammoxidation of propene. Its discovery initiated the development of several other binary oxide catalysts having high selectivities comparable to the bismuth molybdates. For example, a tin-phosphorus catalyst has been developed for the oxidative dehydrogenation of butene to butadiene.[33]

The bismuth molybdate catalysts evolved into multicomponent molybdates, such as one of the Sohio® multicomponent molybdate catalysts patented for the ammoxidation process: direct conversion of propene, oxygen, and ammonia to acrylonitrile.[32] In this reaction, nitrogen is introduced as a functional group:

$$CH_2{=}CH{-}CH_3 + NH_3 + 3/2O_2 \rightarrow CH_2{=}CHCN + 3H_2O \quad (12)$$

Reactions catalyzed by these bismuth molybdates include, in addition to that listed above (Reaction 12):

$$CH_2{=}CH{-}CH_3 + O_2 \rightarrow CH_2{=}CH{-}CH{=}O + 3/2H_2O \quad (13)$$

$$CH_2{=}CHCH_2CH_3 + 1/2O_2 \rightarrow CH_2{=}CH{-}CH{=}CH_2 + H_2O \quad (14)$$

Reaction 13 is also performed commercially with bismuth molybdate catalysts.

In spite of the commercialization of these processes in the 1960s, the basic understanding of the role of the catalyst lagged far behind the technological knowledge of the process. There has been a surge of research activity in the field of catalysis of propene oxidation, which has subsequently been directed to the production of acrylic acid and acrylonitrile as well as acrolein, for there is a close relationship between these products and many instances of the same catalyst serving for the production of all three. Discussion of the production of acrylonitrile as the main reaction product is deferred to the concluding section: here we are concerned with the structural aspects of the Bi_2O_3-MoO_3 system and its role in selective oxidations.

A. Structure of Bismuth Molybdates

Although several research groups have examined the binary-phase system Bi_2O_3-MoO_3, there is general agreement only on the existence of the compounds Bi_2MoO_6, $Bi_2Mo_3O_{12}$, and $Bi_2Mo_2O_9$.

The Bi_2MoO_6 compound is a layered structure with alternating sheets of Bi_2O_2 and MoO_2, the sheets being connected by oxygen layers. The Mo^{6+} ions are in a distorted octahedral surrounding with corner sharing of the octahedra in the layers and the octahedral apex ions pointing to Bi^{3+}-cations. XPS results of Bi_2MoO_6 have shown only Mo^{6+} on the surface at 300 K; heating it to 670 K in vacuum changes it entirely to Mo^{5+} which then gives rise to Mo^{4+} and Mo^{6+} on cooling.[25]

The structure of $Bi_2Mo_3O_{12}$ can be derived from the scheelite structure ($CaWO_4$) by replacing $3Ca^{2+}$ with $2Bi^{3+}$ and a cation vacancy. The MoO_4 tetrahedra form Mo_2O_8 pairs, of which there are two forms.

The structure of $Bi_2Mo_2O_9$ appears as a transition from one type of structure to another. Its most remarkable feature is the presence of rows of oxygen that are only connected to Bi^{3+}. The Van de Elzen and Rieck[34] structure, based on X-ray powder diffraction measurements, can be described as $Bi(Bi_3\phi O_2)(Mo_4O_{16})$, where ϕ stands for a cation vacancy. It is incongruent and only stable between 560 and 660°C. Below this range it dissociates into $Bi_2Mo_3O_{12}$ and Bi_2MoO_6 but IR spectroscopy seems to indicate that it is still present in small amounts, presumably as a surface layer.

A more extensive survey of the structural details has been recently given by Schuit and Gates.[16,35] There is a structural change in Bi coordination from $Bi_2Mo_2O_9$ to $Bi_2Mo_3O_{12}$. In Bi_2MoO_6, Bi^{3+} is sixfold coordinated with two oxygens bound only to bismuth, whereas in $Bi_2Mo_3O_{12}$ all oxygens are shared by bismuth in an eightfold surrounding and molybdenum. Furthermore, there is a decreasing degree of clustering of the Mo-O polyhedra, going from an infinite two-dimensional structure via Mo_4O_{16} to Mo_2O_8. Finally, there is a trend in the concentration of Bi^{3+}-cation vacancies. The ratio (cation vacancie-s/$2Bi^{3+}$) ranges from 0 for Bi_2MoO_6, and 1/2 for $Bi_2Mo_2O_9$ to 1 for $Bi_2Mo_3O_{12}$. A summary of the changes occurring is given in Table 4.

B. Role of Oxygen

In the literature, there is disagreement on the structure of the active 'site' for the catalytically selective and active bismuth molybdates. This is mainly due to the difficulty in preparing stoichiometric, single-phase compounds. It is, therefore, difficult to identify the surface structure of the active 'site(s)' with ill-defined catalysts since the samples are almost nonstoichiometric and/or multiphase. To begin with, we shall discuss here only the last two reactions, 13 and 14, since these are relatively the best understood.

The first point that has to be stressed is that the actual oxidation occurs by oxygen anions

Table 4
STRUCTURE OF SEVERAL BISMUTH MOLYBDATES

Formula	Number of Bi^{3+} of MoO_4 Vacancies	Degree of Clustering of MoO_4
$Bi_2Mo_3O_{12}$ (α-phase)	1	2
$Bi_2Mo_2O_9$ (β-phase)	1/2	4
Bi_2MoO_6	0	Infinite

of the catalyst (Mars-Van Krevelen mechanism). As discussed earlier, this causes a reduction of the catalyst and the only role of gaseous oxygen is to reoxidize the reduced catalyst. Provided the reduction remains small, reoxidation is fast.

Second, the reduction reaction (Equations 15 and 16) formally but not mechanistically consists of a step-wise fission of hydrogen atoms from olefin:

$$\begin{array}{c} CH_2{=}CH{-}CH_2{-}CH_3 \\ \downarrow -H^{\bullet} \\ [CH_2 \;\therefore\; CH \;\therefore\; CH{-}CH_3] \\ \downarrow -H^{\bullet} \\ CH_2{=}CH{-}CH{=}CH_2 \end{array} \tag{15}$$

$$\begin{array}{c} CH_2{=}CH{-}CH_3 \\ \downarrow -H^{\bullet} \\ [CH_2 \;\therefore\; CH \;\therefore\; CH_2] \\ \downarrow -H^{\bullet} \\ \quad\;\; H \\ [CH_2{=}CH{-}\underset{\bullet}{C}\cdot] \\ \downarrow +\text{ O from cat.} \\ \quad\;\; H \\ CH_2{=}CH{-}C{=}O \end{array} \tag{16}$$

For the butadiene formation loss of an α-hydrogen is faster than O-insertion, resulting in diene formation; however, for propene to acrolein, no α-hydrogen atoms are present, and the slower oxygen insertion dominates. The fission of the first C-H starts at the position α to the double bond and the first intermediate is therefore invariably an allyl-radical or ion. The actual mechanistic steps by which these processes occur will be subsequently discussed.

The most remarkable aspect of all these reactions is that none of the products, butadiene or acrolein, show much tendency to become oxidized under the reaction conditions. There are four main schools of thought on this subject as summarized below.

Sohio researchers[32,36,37] assumed that of all the oxygens at the surface in selective bismuth molybdates catalysts, only a few are active enough to give rise to oxidation: they occur in a 'sea' of relatively inert oxygens. This model was later shown to be also valid for oxidations over USb_3O_{10},[38] with the conclusion that the allylic intermediate is a common intermediate.

Another observation has been made by Sleight and Linn[39] using Bi-doped $PbMoO_4$ as the

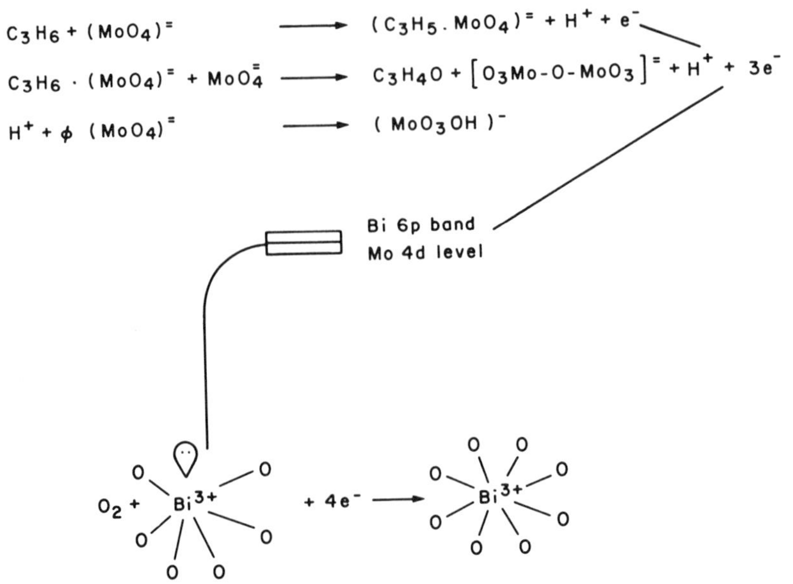

$$C_3H_6 + (MoO_4)^= \longrightarrow (C_3H_5 \cdot MoO_4)^= + H^+ + e^-$$

$$C_3H_6 \cdot (MoO_4)^= + MoO_4^= \longrightarrow C_3H_4O + [O_3Mo\text{-}O\text{-}MoO_3]^= + H^+ + 3e^-$$

$$H^+ + \phi\,(MoO_4)^= \longrightarrow (MoO_3OH)^-$$

Bi 6p band

Mo 4d level

$$O_2 + Bi^{3+}(O)_6 + 4e^- \longrightarrow Bi^{3+}(O)_8$$

FIGURE 7. Reaction diagram to illustrate oxidation of propene over $Pb_{1\text{-}3x}\,Bi_{2x}\,V_x^c\,(MoO_4)$. (From Sleight, A. W. and Linn, W. J., *Ann. N.Y. Acad. Sci.*, 272, 22, 1976. With permission.)

catalyst. Such catalysts are good models to study the reaction since they show all the reactions of the commercial catalysts. $PbMoO_4$ has a scheelite structure and Bi^{3+} can replace Pb^{2+} to some extent to give $Pb_{1\text{-}3x}Bi_{2x}V_x^C(MoO_4)$ with V^C a cation vacancy. They have concluded that the reaction needed both Bi and a cation vacancy. Their hypothesis to explain this observation is given in Figure 7.[40] The reaction starts at two MoO_4 groups that are connected to an empty cation site. As a result, the two MoO_4 groups tend to lose oxygen anions in a step-wise reaction. First, an allyl is formed, π is bonded to one Mo, while H is bonded to an oxygen of a neighboring MoO_4 group. An oxygen is then abstracted from the first MoO_4 group and another H is abstracted. These authors assumed an overlap of the higher unfilled energy levels of Bi^{3+} and Mo^{6+}: if electrons are set free by the reduction, the electrons flow via this conduction band to the Bi cation. This originally occurs in an eightfold surrounding before it donates two oxygens to the two residual MoO_3 groups, thereby adopting a sixfold surrounding. Oxygen molecule now attacks at the Bi position, takes up the electrons, and reestablishes the eight fold surrounding.

The third picture emerges from the work of Matsuura et al.[41-43] These authors found Bi_2MoO_6 to be inactive but it becomes highly active and selective when small amounts of MoO_3 are added. This excess Mo remained in surface layers (as shown by XPS and ISS). In Figure 8, a plot of surface vs. bulk Bi/Mo values for 'multiphase' samples and 'compound' Bi-molybdates is shown. It was concluded that a surface phase, similar to the bulk $Bi_2Mo_2O_9$ phase (as indicated by IR and Raman spectroscopy), existed on the surface phase of bismuth molybdate samples having bulk Bi/Mo values between the stoichiometries 2/1 and 2/3 with the corresponding Bi/Mo surface values near 1.

The last observation described here is that of Keulks and Krenzke,[44,45] who studied the oxidation with oxygen 18 isotope (^{18}O) as the oxidant, the oxygen in the catalyst being ^{16}O. In the absence of propene, there is no exchange between gas-phase oxygen and oxygen in the catalyst structure. However, if a stream of propene and $^{16}O_2$ was led over the catalyst while producing acrolein, and the gas-phase $^{16}O_2$ was abruptly replaced by $^{18}O_2$, the acrolein formed began to show the presence of ^{18}O. This, however, occurred so slowly that they proceeded to assume that the entire amount of the ^{16}O in the catalyst structure had to be

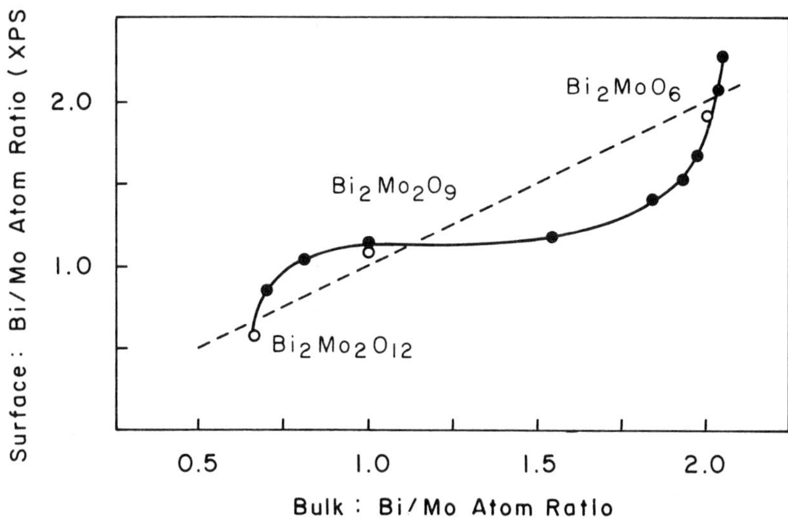

FIGURE 8. Atomic ratio of Bi/Mo at surface from XPS analysis as a function of bulk composition for 'multiphase sample' and compound Bi-molybdates. (From Matsuura, I., Schuit, R., and Hirakawa, K., *J. Catal.*, 63, 152, 1980. With permission.)

replaced by ^{18}O before all the acrolein formed contained only ^{18}O. One concludes that the reoxidation occurs at a site remote from where the interaction with the hydrocarbon took place, probably at a different surface plane. Furthermore, they proposed that the actual diffusion of oxygen through the lattice was rate determining, in disagreement with the results of Schuit and co-workers.[40] These authors had shown that the reduction of the fully oxidized, and the reoxidation of a reduced catalyst were fast enough to be in agreement with this model.

We thus have an interesting but imprecise picture of the catalytic site on bismuth molybdate. It is apparent from these studies that the active ensemble for propene partial oxidation contains both Bi and Mo and perhaps a cation vacancy and may involve a surface defect or domain. It may also be concluded that the bismuth molybdate catalysts are bifunctional — the active ensemble for catalyst reoxidation by gas-phase oxygen is distinct from the active ensemble for propene activation and partial oxidation.

Considering the essential elements in the reaction mode, Sohio researchers[46-48] recently suggested a detailed mechanism which begins with the formation of an allylic intermediate. They have also established the next steps of the sequence. The entire closed sequence of the mechanism is shown in Figure 9.[48] The active and selective site is composed of Bi-Mo pairs, in which the α-hydrogen-abstracting bismutyl (Bi=O) is bonded through a bridging oxygen to the olefin chemisorption/oxygen-insertion Mo-dioxo functionality. Subsequent steps include the formation of O-σ-complex and then acrolein. Therefore, the function of oxygens associated with Bi is to perform α-hydrogen abstraction, while those in Mo polyhedra are sites for olefin chemisorption and O-insertion. To complete the cycle, lattice oxygen is used to form acrolein in the catalyst reduction step, and reoxidation of metal oxide by gaseous oxygen reconstitutes the active oxidant.

Attempts to separate the functions of the components of other antimonate catalysts (e.g., SnO_2 + Sb_2O_3; Fe_2O_3 + Sb_2O_3; and UO_3 + Sb_2O_3), have had success in drawing general analogies to the molybdate and antimonate systems.[48] Sb^{3+} plays a role similar to that of Bi as the α-H abstracting agent, while Sb^{5+} polyhedra serve as the O-insertion component analogous to Mo. However, unlike the molybdates, surface participation accounts for a major portion of catalyst reduction under normal reaction conditions for antimonates, as opposed

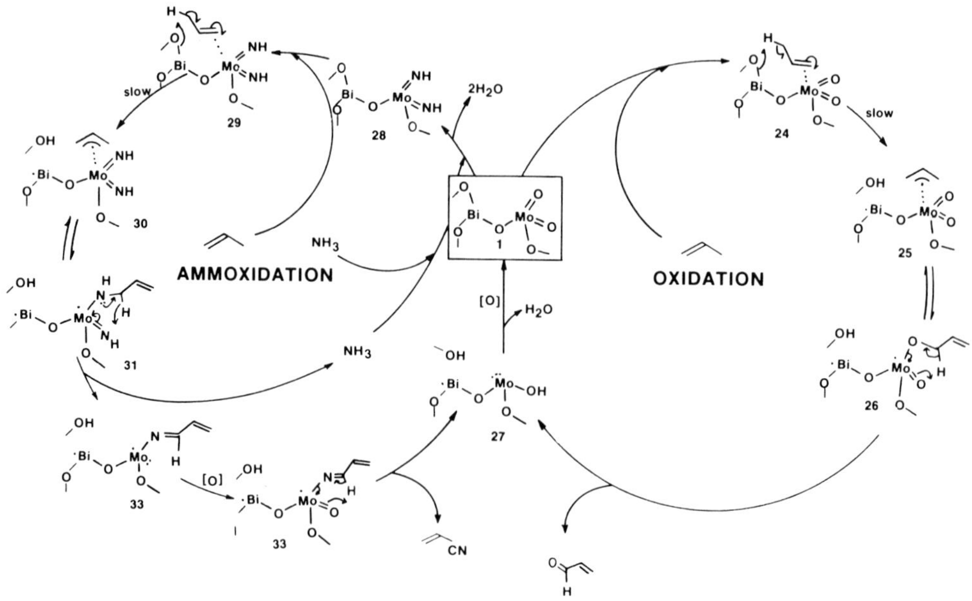

FIGURE 9. Mechanism of selective oxidation and ammoxidation of propene over bismuth molybdates. (From Burrington, J. D., Kartisek, C. T., and Grasselli, R. K., *J. Catal.*, 87, 363, 1984. With permission.)

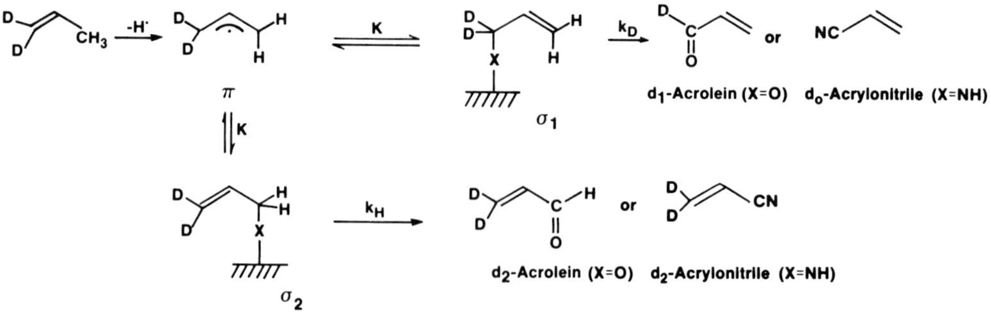

$K(\pi \rightleftharpoons \sigma)$	d_2/d_1-Acrolein	d_2/d_o Acrylonitrile
~1.0 (reversible)	k_H/k_D (1.5-2.5)	1.5-2.5
Very large (reversible)	1.0	1.0
• Bismuth Molybdates	1.63[a] ($Bi_2O_3 \cdot 2MoO_3$)	1.59[b] ($Bi_2O_3 \cdot 3MoO_3$)
• Antimonates	0.96[a] ($Sb_2O_4 \cdot 3SnO_2$)	1.27[b] (USb_3O_{10})

[a] J. J. Portefaix, F. Fiqueras, and M. Forrissier, J. Catal., 63, 307 (1980); 688 K
[b] Ref. (48); 593 K

FIGURE 10. Isotope effect for second hydrogen abstraction in molybdates vs. antimonates. (From Burrington, J. D., Kartisek, C. T., and Grasselli, R. K., *J. Catal.*, 87, 363, 1984. With permission.)

to the bulk participation mechanism which operates for molybdates.[48] An additional distinction is the reversibility of the oxygen insertion for molybdates vs. the irreversibility for antimonates, as measured by the difference in d_2/d_1 acrolein ratios between the two systems using 1, 1-d_2 propene as feed (Figure 10).[48] These key features of the antimonate mechanism are illustrated in Figure 11.

It may be concluded from the considerable effort that has already been expended on the study of multioxide catalysts that those which are effective are either compounds or solid

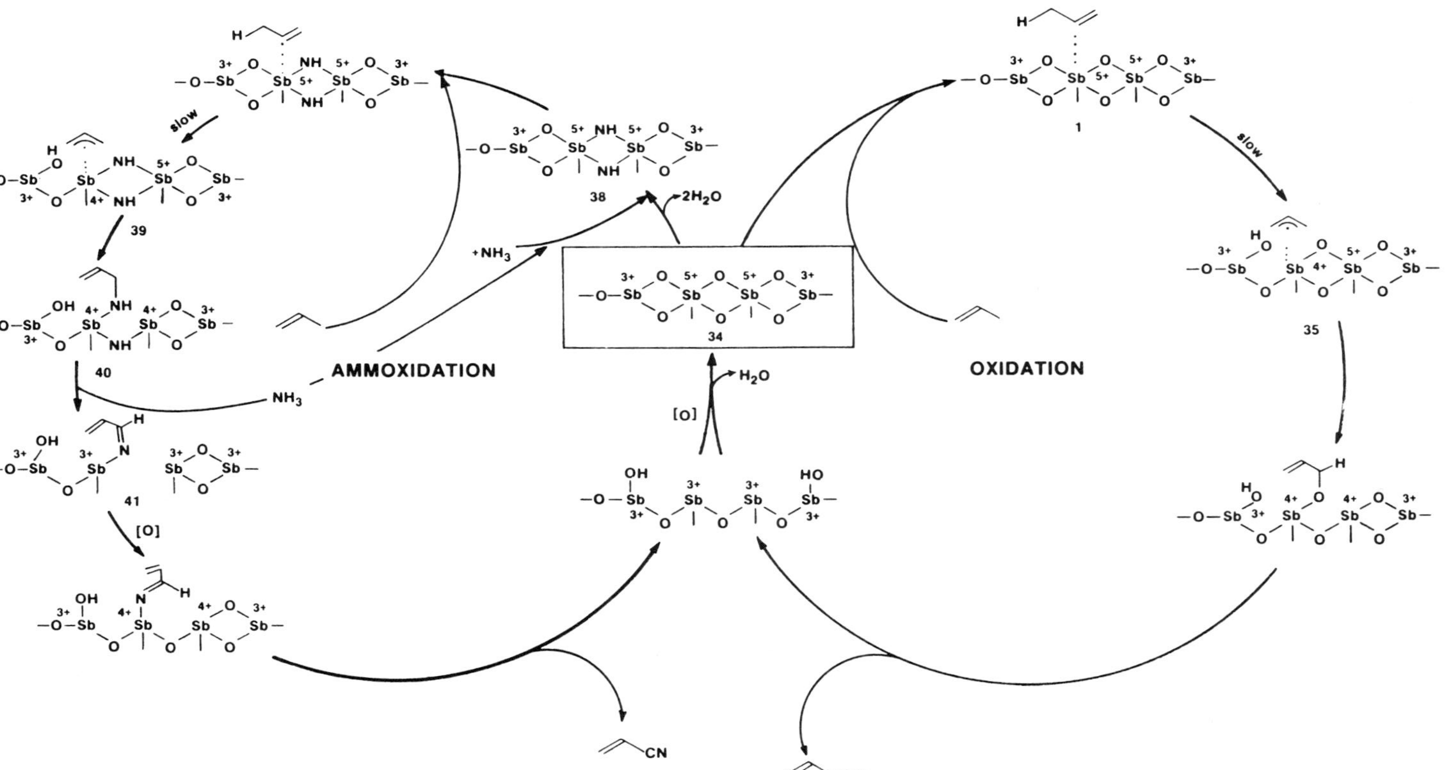

FIGURE 11. Mechanism of selective ammoxidation and oxidation of propene over antimonates. (From Burrington, J. D., Kartisek, C. T., and Grasselli, R. K., *J. Catal.*, 87, 363, 1984. With permission.)

solutions of the component oxides. This appears to be a necessary condition. Although combination of the oxides in either a random or stoichiometric manner has not been observed in all cases, it can be reasonably assumed to be a general rule, as the pronounced synergism of the effective systems could hardly arise without molecular intimacy of the constituents.

V. AMMOXIDATION OF PROPENE TO ACRYLONITRILE

The ammoxidation reaction, as mentioned previously, proceeds according to

$$CH_2{=}CH{-}CH_3 + NH_3 + 3/2O \rightarrow CH_2{=}CHCN + 3H_2O \tag{12}$$

The reaction is highly exothermic, and therefore, it is commonly carried out in a fluidized-bed reactor to facilitate rapid heat removal. Other products are acetonitrile, hydrogen cyanide, carbon dioxide, and carbon monoxide, whereas acrolein is only a trace product.

In the absence of propene, the oxidation of ammonia also takes place according to

$$2NH_3 + 3/2O_2 \rightarrow N_2 + 3H_2O \tag{17}$$

The catalyst is composed of a variety of elements like nickel, cobalt, iron, manganese, potassium, and phosphorus but always contains bismuth and molybdenum. Today, this catalytic process accounts for virtually all of the eight billion lb annual production of the world.

The first generation catalyst, $Bi_9PMo_{12}O_{52}$ on SiO_2, produced 65% per pass propene conversion to acrylonitrile. Yield of useful products has greatly increased with newer generation catalyst systems. For example, $Co_6^{2+}Ni_2^{2+}Fe_3^{3+}Bi^{3+}$ on 50 wt% SiO_2 with some P and K (referred to as an MCM [multicomponent molybdate]). It is generally accepted that bismuth molybdate forms the active phase in these multifunctional, multicomponent molybdate catalysts. The additional components function to stabilize the catalysts for a long reactor lifetime with minimum deactivation (in the case of the other transition metal molybdates) resulting mainly from MoO_3 transition, supply the necessary mechanical strength for fluidized-bed operation as bismuth molybdate alone is mechanically weak, and increase catalytic activity to allow processing at lower temperatures (the typical operating temperature range is 673 to 773 K). However, the exact function and disposition of these new components in MCM remain unclear. To simplify matters, the following discussion is restricted to the main components only.

Broadly speaking, there is great similarity in the overall features of the oxidation and ammoxidation of propene. Generally, the same catalyst can be used for either reaction. The reduction of the catalyst and oxidation on its surface have similar characteristics. The details of reaction mechanism are now discussed.

A. Kinetics and Mechanism

As for oxidation, the bismuth molybdate-catalyzed rate of ammoxidation is first order with respect to propene. Likewise, activation energies for the formation of acrylonitriles are similar to those for oxidation (~96 kJ/mol).

As shown in Figures 9 and 11, the mechanistic steps for ammoxidation are the same as those for oxidation, except that NH_3 activation occurs prior to α-H abstraction by formation of Mo=NH groups which serve as the nitrogen insertion components. Also, since ammoxidation is a six-electron oxidation (vs. a four for oxidation), an intermediate reoxidation step probably occurs prior to acrylonitrile desorption.

Both oxygens of the Mo-dioxo group are replaced by NH, as measured by the dependence of acrolein/acrylonitrile (AN) ratio on the ammonia/propene feed ratio, which is known to be kinetically related to the number of ammonia molecules activated per site in the acrylonitrile-forming cycle.[47] For typical turnover rates, the AN/acrolein product ratio is a linear function of $(NH_3)^2/C_3H_6$, corresponding to the activation of ammonia in pairs, but with the formation of only one N-inserting species.[48]

B. Role of Iron in Multicomponent Molybdate

The function of iron is of considerable interest because it is the main trivalent cation that is used as a promoter for bismuth molybdates. This is also the most studied component outside the patent literature.

Batist et al.[52] were the first to report the presence of molybdates containing both Fe and Bi. Sleight and Jeitschko[53] established the structure of the compound $Bi_3(FeO_4)(MoO_4)_2$ and Lo Jacono et al.[54] gave some suggestions for the existence of still another compound with a Bi/Fe ratio of 1. In an inert atmosphere, it is possible that $Fe_2(MoO_4)_3$ may oxidize Bi metal to Bi(III) molybdate under formation of Fe(II) molybdate. Therefore, the interaction between Fe and Bi occurring in a ternary compound might be of importance.

Based on the observed kinetics, Brazdil et al.[49] concluded that iron in the $Bi_3FeMo_2O_{12}$ scheelite structure promotes the rapid exchange oxygen between the reoxidation sites, and the bulk by means of interlinking Fe^{3+}/Fe^{2+} redox couples in the partially reduced catalyst. The oxygen mobility in the MCM system is enhanced by the presence of shear structures in the lattice.

An alternative is that the reaction

$$Bi^{n+} + Fe^{3+} \rightleftharpoons Fe^{2+} + Bi^{(n+1)+} \tag{18}$$

is an equilibrium reaction. This would enable the bismuth to perform its usual function while preventing a large concentration of Bi^{n+}. If $n = 0$, one needs three Fe^{3+} cations for the conversion to Bi^{2+}, which is the same as the value for the Fe/Bi ratio in a typical commercial catalyst. Usually, the Bi atoms are rapidly reconverted to Bi_2O_3 by oxygen, but some of them would tend to form Bi metal. It has been shown that the semiconductivity of the MCM catalyst during pulse operation is 100 times less than in the pure Bi molybdate.[55] Schuit and Gates[35] therefore concluded that the probability of forming MoO_3 is strongly decreased since this component is somewhat volatile at the reaction temperature; hence, the catalyst is expected to have longer life.

In summary, it implies that the multicomponent catalyst is unique in the particularly facile pathways of oxygen transfer and site reconstruction; electron exchange is available to it.

REFERENCES

1. **Szonyi, G.,** *Adv. Chem. Ser.*, 70, 53, 1968.
2. **Henry, P. M.,** *Adv. Chem. Ser.*, 70, 126, 1968.
3. **Stern, E. W.,** *Catal. Rev.*, 1, 73, 1967.
4. **Falbe, J.,** *Carbon Monoxide in Organic Synthesis*, Springer-Verlag, Berlin, 1970.
5. **Roth, J. F.,** *Platinum Met. Rev.*, 19, 12, 1975.
6. **Lyons, J. E.,** in *Applied Industrial Catalysis*, Vol. 3, Leach, B. E., Ed., Academic Press, New York, 1983, chap. 6.
7. **Gould, E. S. and Rado, M.,** *J. Catal.*, 13, 238, 1969.
8. **Neuberg, H. J., Phillips, M. J., and Graydon, W. F.,** *J. Catal.*, 38, 33, 1975.
9. **Sharma, R. K. and Srivastava, R. D.,** *J. Catal.*, 39, 315, 1975.
10. **Agarwal, A. K. and Srivastava, R. D.,** *J. Catal.*, 45, 86, 1976.
11. **Krishna, L. V. G., Rao, M. S., and Srivastava, R. D.,** *J. Catal.*, 49, 109, 1977.

12. **Sheldon, R. and Kochi, J.,** *Metal-Catalyzed Oxidation of Organic Compound,* Academic Press, New York, 1981.
13. **Henry, P.,** *Palladium Catalyzed Oxidation of Hydrocarbon,* Kluwer Academic, Boston, 1980.
14. **Evnin, A. B., Rabo, J. A., and Kasai, P. H.,** *J. Catal.,* 30, 109, 1973.
15. **Seoane, J. L., Boutry, P. and Montarnal, R.,** *J. Catal.,* 63, 182, 1980.
16. **Gates, B. C., Katzer, J. R., and Schuit, G. C. A.,** *Chemistry of Catalytic Processes,* McGraw-Hill, New York, 1979., chap. 4.
17. **Sharma, R. K., Rai, K. N., and Srivastava, R. D.,** *J. Catal.,* 63, 271, 1980.
18. **Sharma, R. K. and Srivastava, R. D.,** *J. Catal.,* 65, 481, 1980.
19. **Mars, P. and Van Krevelen, D. W.,** *Chem. Eng. Sci. Suppl.,* 3, 41, 1954.
20. **Agarawal, D. C., Nigam, P. C., and Srivastava, R. D.,** *J. Catal.,* 55, 1, 1978.
21. **Sharma, R. K. and Srivastava, R. D.,** *AIChE J.,* 27, 41, 1981.
22. **Sharma, R. K. and Srivastava, R. D.,** *AIChE J.,* 28, 855, 1981.
23. **Srivastava, R. D., Stiles, A., and Jones, G. A.,** *J. Catal.,* 77, 192, 1982.
24. **Srivastava, R. D. and Jones, G. A.,** unpublished results.
25. **Jagannathan, K., Srinivasan, A., and Rao, C. N. R.,** *J. Catal.,* 69, 418, 1981.
26. **Andersson, S. L. T. and Jaras, S.,** *J. Catal.,* 64, 51, 1980.
27. **Andersson, S. L. T.,** *J. Chem. Soc., Faraday Trans. 1,* 75, 1356, 1979.
28. **Anderson, A.,** *J. Catal.,* 69, 465, 1981.
29. **Anderson, A. and Lundin, S. T.,** *J. Catal.,* 65, 9, 1980.
30. **Anderson, A. and Lundin, S. T.,** *J. Catal.,* 58, 383, 1979.
31. **Callahan, J. L., Foreman, R. W., and Veatch, F.,** U.S. Patent 3,044,966, 1962.
32. **Callahan, J. L., Grasselli, R. K., Milberger, E. C., and Strecker, H. A.,** *Ind. Eng. Chem. Prod. Res. Dev.,* 9, 134, 1970.
33. **Pitzer, E. W.,** *Ind. Eng. Chem. Prod. Res. Dev.,* 11, 299, 1972.
34. **Van den Elzen, A. F. and Rieck, G. D.,** *Mat. Res. Bull.,* 10, 1163, 1975.
35. **Schuit, G. C. A. and Gates, B. C.,** *Chem. Technol.,* 693, November, 1983.
36. **Grasselli, R. K. and Burrington, J. D.,** *Adv. Catal.,* 30, 133, 1981.
37. **Grasselli, R. K.,** in *Heterogenous Catalysis: Selected American Histories,* Davis, B. H. and Hettinger, W. P., Eds., ACS Symposium Ser. 222, Washington, D.C., 1983, chap. 25.
38. **Grasselli, R. K. and Suresh, D. D.,** *J. Catal.,* 25, 273, 1972.
39. **Sleight, A. W. and Linn, W. J.,** *Ann. N.Y. Acad. Sci.,* 272, 22, 1976.
40. **Schuit, G. C. A.,** Catalysis by Oxides and Sulfides, in Symposium on Quantum Chemistry, Upsula, 1977, preprint.
41. **Matsuura, I. and Schuit, G. C. A.,** *J. Catal.,* 20, 19, 1971.
42. **Matsuura, I. and Schuit, G. C. A.,** *J. Catal.,* 25, 314, 1972.
43. **Matsuura, I., Schuit, R., and Hirakawa, K.,** *J. Catal.,* 63, 152, 1980.
44. **Keulks, G. W.,** *J. Catal.,* 19, 232, 1970.
45. **Keulks, G. W. and Krenzke, L. D.,** Proc. 6th Int. Congr. Catal., London, 1976, preprint B-20.
46. **Burrington, J. D., Kartisek, C. T., and Grasselli, R. K.,** *J. Catal.,* 63, 235, 1980.
47. **Burrington, J. D., Kartisek, C. T., and Grasselli, R. K.,** *J. Catal.,* 81, 489, 1983.
48. **Burrington, J. D., Kartisek, C. T., and Grasselli, R. K.,** *J. Catal.,* 87, 363, 1984.
49. **Brazdil, J. F., Suresh, D. D., and Grasselli, R. K.,** *J. Catal.,* 66, 347, 1980.
50. **Wragg, R. D., Ashmore, P. G., and Hockey, J. A.,** *J. Catal.,* 31, 293, 1973.
51. **Cathala, M. and Germain, J. E.,** *Bull. Soc. Chim. Fr.,* 2167, 2174, 1971.
52. **Batist, P. A., van de Moesdijk, C. G. M., Matsuura, I., and Schuit, G. C. A.,** *J. Catal.,* 20, 40, 1971.
53. **Sleight, A. W. and Jeitschko, W.,** *Mat. Res. Bull.,* 9, 951, 1974.
54. **Lo Jacono, M., Notermann, T., and Keulks, G. W.,** *J. Catal.,* 40, 19, 1975.
55. **van Oeffellen, D. A. G.,** Ph.D. thesis, University of Technology, Eindhoven, The Netherlands, 1978.
56. **Hodnett, B. K.,** *Cat. Rev. Sci. Eng.,* 27, 273, 1985.

Chapter 4

EPOXIDATION OF ETHENE TO ETHENE OXIDE

I. GENERAL

Ethene oxide (EO) is one of the most versatile chemical intermediates used for the manufacture of ethene glycol, detergents, polyurethane foams, and synthetic fibers. EO itself is used as a fumigant for insect control on grain. It was first prepared by Wurtz,[1] by reacting ethene, chlorohydrin, and potassium hydroxide. Wurtz, in 1863, reported the lack of success in obtaining ethene oxide via direct oxidation.

In 1931, Lefort[2,3] succeeded in preparing ethene oxide from ethene and oxygen over a silver catalyst. The older 'chlorohydrin process', involving the reaction of ethene with hypochlorous acid

$$CH_2{=}CH_2 + HOCl \rightarrow CH_2Cl{-}CH_2OH \rightarrow \overset{\;\;\;O\;\;\;}{CH_2 \diagup \!\!\!\! - \!\!\!\! \diagdown CH_2} + HCl \qquad (1)$$

was inefficient in terms of yield, partly because it consists of two characteristic reaction steps. The driving force behind finding a direct oxidation route was the problem of removing and selling the chlorinated hydrocarbon by-products and the need for large volumes of chlorine required in the chlorohydrin process. For a good discussion of the chlorohydrin process, the reader is referred to a recent review by Berty.[4]

Today, ethene oxide (also called ethene epoxide) is almost exclusively produced by vapor-phase oxidation of ethene over selective silver catalysts. The reaction network is shown below:

$$\begin{array}{ccc} C_2H_4 & \xrightarrow[r_1]{+1/2O_2} & C_2H_4O \\ +3O_2 \searrow r_2 & & r_3 \swarrow +5/2O_2 \\ & 2(CO_2 + H_2O) & \end{array} \qquad (2)$$

This epoxidation reaction is the simplest example of a kinetically controlled, selective heterogeneous catalytic reaction. Metals other than silver generally produce CO_2 and H_2O, the thermodynamically preferred product. There are several points in the triangular network which can be influenced by promoters, supports, etc. A conversion selectivity curve for laboratory oxidation is shown in Figure 1. A striking characteristic of ethene oxidation is that the limiting selectivity with moderated silver is insensitive to conversions and temperatures. As such, this reaction is of fundamental importance to catalytic and surface scientists, since silver is unique in its ability to synthesize EO with high selectivity, minimizing the formation of CO_2 and water, and has therefore been the subject of numerous studies.

It has been stated that the maximum selectivity of the process is 6/7 (86%). This figure is based on the hypothesis that only molecularly adsorbed O_2 yields epoxide by reaction with ethene and that only one molecule takes part in this reaction. The selective reaction is

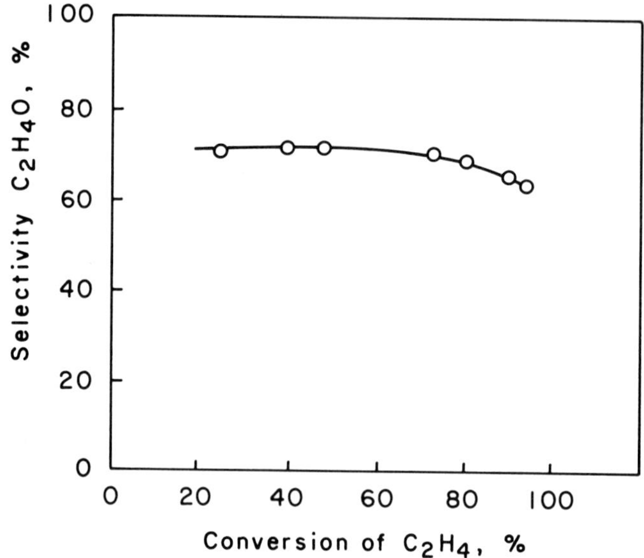

FIGURE 1. Selectivity-conversion curve for ethene oxidation.[5]

$$CH_2{=}CH_2 + O_2(a) \rightarrow \overset{\quad O}{CH_2 - CH_2} + O(a) \tag{3}$$

The only by-product formed is CO_2 and H_2O:

$$CH_2{=}CH_2 + 6O(a) \rightarrow 2CO_2 + 2H_2O \tag{4}$$

Although Kilty and Sachtler[6] reported experimental data in favor of this limiting selectivity hypothesis, such mechanistic arguments based upon reaction selectivity are questionable due to the fact that molecularly adsorbed oxygen likely represents a small fraction of the total oxygen adsorbed and much larger amounts of atomic oxygen are present than would be produced by Reaction 3.[7] Nevertheless, the adsorption state of oxygen has played a central role in the development of recent ideas on the mechanism of selective oxidation of ethene.

Certain alkali metal-promoted catalysts have been found to display selectivities higher than the limiting selectivity. In industrial processes, chlorinated hydrocarbons are added in trace quantity to the reactant feed in order to improve catalyst selectivity for ethene oxide.

In the past 10 years the number of papers devoted to catalysis and surface science has increased and this is reflected in the material in this chapter and its arrangement. The single-crystal studies have provided considerable new insight into the reaction pathway, the structural sensitivity of the silver catalysts, and the role of chlorine adatoms in promoting catalyst selectivity. The fundamental question which remains unanswered for this process is the identity of the surface oxygen species responsible for the epoxidation reaction and combustion reactions. The following sections illustrate the direction in which applications are proceeding. It is hoped that the references given to reviews[5-12] will be of value to readers.

This chapter proceeds with an introductory description of the industrial processes, catalysts, and an overview of the reaction kinetics followed by a detailed account of the state of the various adsorbed species, and especially that of oxygen. The concluding section focuses on the mechanism of epoxidation.

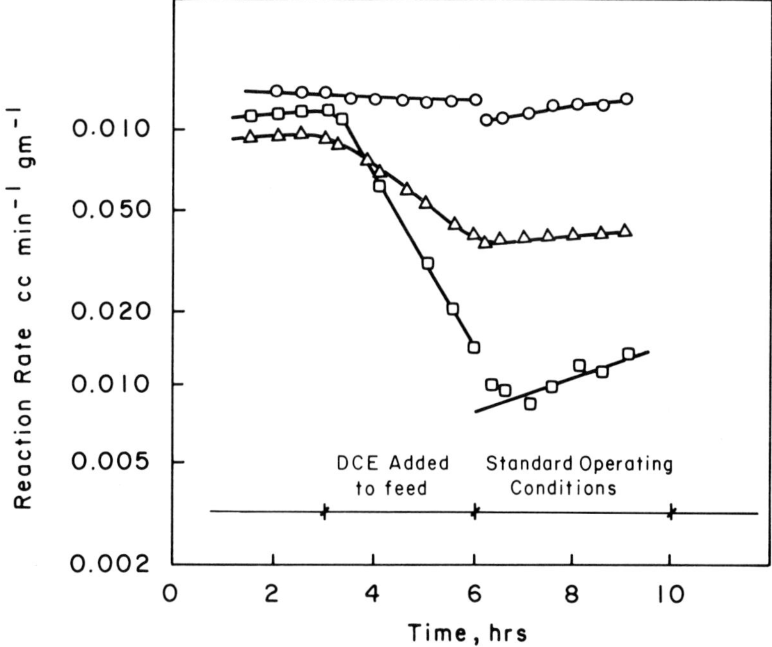

FIGURE 2. Transient response of the rate of ethene oxide formation to dichloroethane addition: (○) 160 ppb DCE, (□) 320 ppb DCE (△) 640 ppb DCE. (Reprinted with permission from Metcalf, D. L. and Harriot, P., *Ind. Eng. Chem. Process Des. Dev.*, 11, 478, Copyright 1972, American Chemical Society.)

II. THE CATALYST AND THE PROCESS

The epoxidation process shows a good deal of variation, probably due to the complexity of the system. In general, ethene and air (ratio or air/ethene $\approx$ 7:8) or oxygen is passed over a silver-based catalyst at approximately 520 to 600 K under 10 to 50 atm pressure.

Plants are usually operated on a recycle system. Due to the inefficiency of CO_2 removal, the recycle contains not only ethene and oxygen, but also a high level of CO_2 (approximately 1%) which does interfere with the synthesis process and is generally detrimental. Also, the production of CO_2 releases 13 times as much heat as the production of ethene oxide, the selectivity must be closely controlled to protect the catalyst from heat damage. The choice of operating conditions is further limited by the fact that C_2H_4-O_2 is an explosive mixture, and also to the effect that ethene oxide polymerizes at the synthesis conditions which must be avoided. Above 570 K, the oxidation becomes rapid, hot spots are formed in the catalyst bed, there is temperature runaway, and complete combustion occurs. The effluent from the Dowtherm-cooled reactor is absorbed in water and subsequently steam stripped for ethene oxide.

The most common catalyst is silver (8 to 12 wt%) supported on α-Al_2O_3, of 1 m^2 g^{-1}-specific surface area, promoted up to 0.35% concentrations of lithium, sodium, and/or potassium, as well as by a doping with up to 0.025% cesium. Various interpretations of the specific action of these additives are reported in the literature.[5,13,14] The use of a silicon-free alpha alumina support for such a catalyst increases the service life of the catalyst.[15]

Trace amounts (1 to 12 ppm) of chlorinated organic compounds (dichloroethane or vinyl chloride) are usually added to the feed. The organic chlorine plays a major role in the synthesis, as discussed later in the text, since it improves the selectivity of the catalyst by inhibiting the total oxidation of ethene to CO_2 and H_2O. The effect of adding several different concentrations of dichloroethane is illustrated in Figure 2.

Table 1
KINETIC PARAMETERS FOR EPOXIDATION AND FOR TOTAL OXIDATION OVER SILVER CATALYSTS

Reaction 1			Reaction 2			
E_1 (kJ/mol)	$C_2H_4^n$	O_2^n	E_2 (kJ/mol)	$C_2H_4^n$	O_2^n	Ref.
67	1	0	67	1	0	30
50	0	1	63	0	1	117
81	0.3	0.7				118
82	0	1	138			119
36	0.5	0	31	0.5	0	120
	0	1		0	1	121
	1	1.5		1	2	122
	1	1.5		1	1.5	67
	−0.03	0.91		−0.2	1.1	123
$E_1 = E_2$	1	0.5	$E_2 = E_1$	1	0.5	32
105	1	0.5	99	1	0.5	124
77	0.7	1	63	0.7	1	125

The silver catalyst used in a large ethene oxidation plant is typically contained in several thousand pipes. When this catalyst loses its activity and selectivity after about 5 years of use, it can either be replaced or regenerated. Moreover, silver is lost, both during replacement and in recovery of silver from the spent catalyst, and has been a constant concern of the industry. Various procedures for regenerating silver catalysts with organic solutions of differing concentrations of calcium and cesium salt have been recently reported.[17,18]

In the past 5 years the number of patents devoted to improve yields of ethene oxide has increased. Special treatment is introduced for reducing the silver salt to catalytically active metallic silver.[19-23] Boron,[24] thallium,[25,26] and antimony[27] have also been shown to improve the selectivity, and it is claimed that these elements tend to be retained by the catalyst. Silver silicate has also been utilized as the active material in an improved ethene oxidation catalyst.[28]

III. KINETIC DATA

The kinetics of silver-catalyzed ethene oxidation have been studied using a differential as well as an integral reactor. There is a general consensus that the overall kinetics can be described by means of a triangular scheme as described in the preceding section (Equation 2). The values of the parameters in the form

$$r_1 = k_{01}\, p_{C_2H_4}^{n}\, p_{O_2}^{n} \exp(-E_1/RT) \tag{5a}$$

and

$$r_2 = k_{02}\, p_{C_2H_4}^{n}\, p_{O_2}^{n} \exp(-E_2/RT) \tag{5b}$$

for selected references are presented in Table 1.[10]

There is a general agreement that two parallel reactions are coupled with each other. It is also known that the further oxidation of the product ($C_2H_4O + 5/2O_2 \rightarrow 2CO_2 + 2H_2O$) plays an important role in determining the selectivity in industrial processes. For example,

Dettwiler et al.[125] found that r_3 was large for their pumice-supported silver catalyst, k_3 exceeding both k_1 and k_2 by more than an order of magnitude. However, most of the above-reported kinetic data have been obtained under the conditions where the rate of consecutive reaction was low.

Findings by different researchers are in general agreement as far as the facts that reactants enhance the reaction and all three products hinder it. The effects of total pressure and temperature on yields and selectivity are also common to all work, but the mechanisms and resulting rate expressions show no more than similar forms, and often not even that.

Each mechanism fits gathered data from one somehow unique catalyst. In addition, differences in the reported kinetic equations depend on the type of chemisorbed oxygen present at the surface. It is likely that rate equations discussed can only be applied to a narrow range of experimental conditions owing to the fact that different conditions produce different surface situations. Since the experimental data appear so conflicting, these schemes will not be discussed here. An intrinsic knowledge of the basic catalytic mechanisms is required. The pertinent information on kinetics at high pressures is proprietary.

IV. INTERACTION OF ETHENE AND OXYGEN: AN OVERVIEW

Worbs[29] proposed the first mechanism for silver-catalyzed ethene epoxidation, and suggested that it was diatomic oxygen that reacted with ethene to form EO, whereas oxygen atoms were responsible for the combustion reaction. Voge and Adams,[5] and later Kilty and Sachtler,[6] extended this theory to say that after the reaction of diatomic oxygen with ethene, the remaining oxygen atoms on the surface undergo full oxidation, the selectivity thus being limited to the classical value of 6/7.

On the other hand, several workers[30,31] have proposed reaction schemes in which the roles of these two oxygen species are reversed. Twigg[30] showed that only oxygen is chemisorbed on silver, both as atoms and molecules, and if the ethene reacts with an adsorbed oxygen atom ethene oxide is formed, otherwise, CO_2 results. Hayes[31] used an approach similar to Twigg's using oxygen ions in place of atoms and molecules.

Others[32-34] have suggested that only a single oxygen species may be involved in both reactions. Gerei et al.[33] have used infrared spectroscopy to determine the adsorbed species formed when ethene is adsorbed on a silver surface (Ag/SiO_2) partially covered by oxygen at 363 K. Subsequent IR spectra showed a band at 870 cm^{-1}, which was attributed to the O–O vibration of an organic peroxide, $CH_2–CH_2–O–O–Ag$. Heating the catalyst to 383 K resulted in a different IR spectrum. Similar structures were observed by these authors when EO was adsorbed on clean silver indicating that adsorption is accompanied by a rupture of the epoxide ring. Kilty et al.[35] confirmed such a structure by examination of the isotopic shifts produced when $^{18}O_2$ and its mixture with $^{16}O_2$ were used.

Force and Bell[32] have also used IR effectively for identifying the structures present on the surface of a silver catalyst during the epoxidation. They carried a detailed study under full reaction conditions at a temperature of 493 K. No bands could unambiguously be ascribed to peroxidic species. This, in conjunction with their kinetic measurements, led the authors to conclude that (1) O^- ions produced on the silver surface are the primary oxidizing agent for both processes, and (2) oxidation proceeds through both gaseous and adsorbed ethene, the former yielding primarily ethene oxide and the latter yielding the combustion products. It may, however, be noted that the postulation of O^- as the sole form of adsorbed oxygen participating in the oxidation mechanism is in conflict with the interpretation given by Kilty and Sachtler.[6]

Evidence which supports the mechanism that $O_2(a)$ are the active species for epoxidation has been made by Herzog,[36] who made the use of nitrous oxide at low pressures as an oxygen atom source for oxidation of ethene and observed only the combustion reaction. At

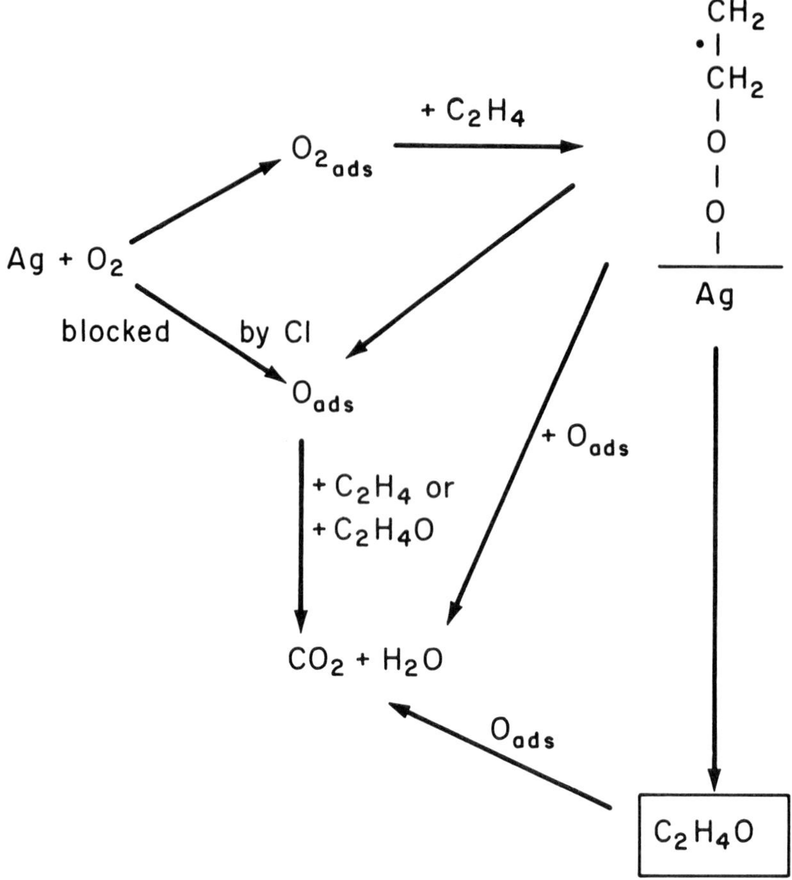

FIGURE 3. Worbs-Sachtler mechanism.[12]

higher temperatures (where N_2O is known to decompose into N_2 and O_2),[37] the epoxidation was the primary reaction observed. This result has also been confirmed by others.[39,40]

Recent results, concentrated upon mechanistic aspects of the reaction, have shown that the situation is rather more complicated than that represented by the mechanism of Worbs. The work of Cant and Hall[34] has shown that the epoxide products are nearly equilibrated. It was also shown by means of isotopes that some rotation about the original C=C bond occurs during epoxidation. A radical reaction mechanism was suggested. This is in contrast to the concerted electrophilic addition usually proposed.[6] Egashira et al.[41] have also shown that the extent of cis-trans equilibration in the epoxide may be related to the oxidation state of silver, with more highly oxidized catalysts producing less equilibrium. The strength of the silver-ethene bond is also known to increase as the silver becomes more electron deficient.[42-45] The results of Egashira thus suggest that the promoters on the silver catalyst may influence the binding of ethene as well as of oxygen to the surface.

Cant and Hall[34] also showed that the selectivity for epoxidation is higher for C_2D_4 than that for C_2H_4.[34] Sachtler and co-workers[10,46] suggested that this result implies that along with the necessity of ethene to combine with surface diatomic oxygen for the production of ethene oxide, the O–O–CH_2–CH_2 complex also has a finite chance of decomposing into carbon dioxide and water (Figure 3[12]). Van Santen et al.[46] interpreted the data of Cant and Hall in terms of a branched scheme:

$$O_2(a) \begin{cases} \xrightarrow[p]{+7/6C_2H_4} C_2H_4O + 1/3(CO_2 + H_2O) \\ \xrightarrow[(1-p)]{+1/3C_2H_4} 2/3(CO_2 + H_2O) \end{cases}$$

$$2O(a) \xrightarrow{+1/3C_2H_4} 2/3(CO_2 + H_2O) \qquad (6)$$

in which the isotope effect is contained in changes in the branching probability, p. According to this mechanism, an increased ratio of O_2(a) to O(a) on the catalyst surface should increase the observed isotope effect.

From the studies of the stoichiometric reaction of deuterated ethene with oxygen adsorbed on silver powder, Shell researchers[47] found that adsorbed atomic oxygen can react to give epoxide, provided that subsurface oxygen is present. More recently, the same group (Van Santen and De Groot[48]) concluded that ethene reacts with atomic oxygen to give epoxide either directly or indirectly after recombination of the oxygen atoms. These authors state that there is, in principle, no reason why the epoxidation selectivity should be limited to 6/7.

From these foregoing discussions, it is apparent that this catalytic reaction system is a complex one. Spectroscopic and chemical data indicated the presence of oxygen species with different charges. Combustion occurs via nucleophilic attack of oxygen on the hydrogens. Epoxidation would appear to require electrophilic attack (at carbon) either concerted, or end on. One of the appeals of the molecular oxygen mechanism is that it is somewhat analogous to known homogeneous (concerted) epoxidations with organic hydroperoxides; though on close examination the orientation and charge distribution on O_2(a) do not appear right. This thrusts the discussion of adsorbed oxygen in the context of determining which species can fulfill the chemical roles that appear to be required.

The following sections will attempt to present the results of the work, which have been obtained by separating out different components of this epoxidation process. Discussion on the adsorption of oxygen on silver follows.

V. ADSORPTION OF OXYGEN ON UNPROMOTED SILVER: IN-DEPTH REVIEW

There have been two general approaches to the study of oxygen interaction with silver surfaces. First, there is the classical approach where coverages and heats of adsorption and desorption have been reported over a range of oxygen pressure (10^{-3} to 300 torr), the 'high pressure' studies. The second approach involves the application of modern surface science techniques to well-defined surfaces and includes mainly the interaction of oxygen with the clean metal in single crystal under ultrahigh vacuum (UHV) conditions. Many of them are at variance with each other, but several common observations may be found in these studies which illuminate the silver-oxygen interaction.

A. Historical Overview and 'High Pressure' Studies

The interaction between oxygen and silver has received much attention for over 50 years. Prior to the development of modern surface science techniques, these studies were conducted at high pressures with more poorly defined materials than the recent studies. The earlier work is reviewed here as it does anticipate some of the observations made in surface studies.

Most of the investigators have been concerned with adsorption and desorption phenomena on silver powders and films. Of the 'high pressure' studies published so far on the adsorption

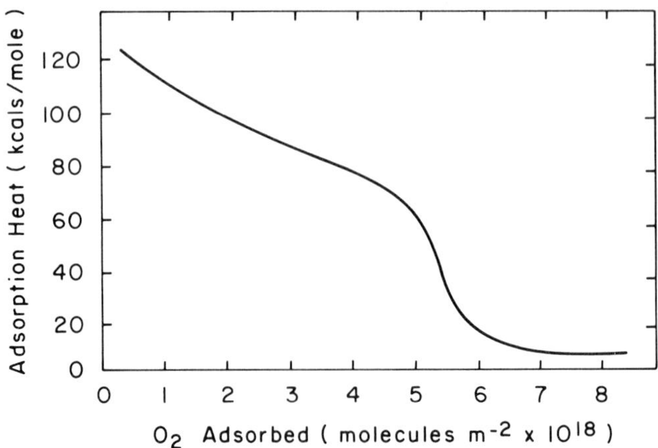

FIGURE 4. Data for the coverage dependence of the heat of adsorption of oxygen on silver powder.[65]

of oxygen on silver powders,[49-51] wires,[52] and polycrystalline films,[53,54] those by Czanderna and co-workers are particularly comprehensive. These authors have examined oxygen adsorption and desorption,[51,55-58] pretreatment effects,[59,60] and interactions on silver field emission tips.[61] The rate of oxygen adsorption and desorption was also measured volumetrically by Smeltzer et al.[62] and manometrically by Kilty et al.[35] As shown by Czanderna,[61] oxygen adsorption and the characteristics of the oxygen adlayer on silver are rather complex; however, several common features were found in these studies;[35,62] they all observed one or more breaks in their kinetic curves, thereby suggesting several adsorbed states of oxygen on silver.

The initial heat of adsorption of oxygen on prereduced silver metal is high (≈336 kJ/mol).[64] As the oxygen coverage increases, the heat of adsorption steadily decreases to 76 to 105 kJ/mol at a coverage of about 0.33 and then remains constant. The activation energy for oxygen adsorption is very low (≤13 kJ/mol). This initial adsorption was considered dissociative.[61] The second adsorption process was considered nondissociative and it occurred with an activation energy of approximately 33 kJ/mol. Czanderna[51] also reported the third process to occur at higher temperatures with an activation energy of 92 kJ/mol, which he ascribed to surface diffusion of oxygen atoms. Kilty et al.[35] also reported three adsorption processes: an 'almost nonactivated' dissociative adsorption, a nondissociative adsorption with an activation energy of 33 kJ/mol, and a dissociative adsorption with an activation energy of 60 kJ/mol. The existence of multiple activation energies of adsorption is also supported by the calorimetric data of Ostrovskii and Temkin.[65] Figure 4 presents their data, although quantitatively inaccurate, and at least shows two breaks, thereby suggesting three adsorbed oxygen species.

At high coverages, greater than at least one half monolayer, Czanderna[51] observed weakly bound oxygen which was assumed to be O_2^-(a). Smeltzer et al.[62] also observed a break in their isothermal adsorption vs. time profile which was attributed to a transition from dissociative to nondissociative adsorption.

Later studies by Czanderna et al.[56,58] showed similar results for oxygen desorption. The oxygen desorption spectra obtained from silver powder[56] and polycrystalline filaments[66] show two distinct peaks at 460 and 650 K. Both desorption processes are second order on oxygen and desorb with activation energies of 113 and 155 kJ/mol.

There was considerable contradiction as to the existence of adsorbed molecular oxygen on the surface,[9,67] though the positive evidence of molecular adsorption was provided by

Clarkson and Cirillo[38] in 1974, who detected O_2^- species using ESR. However, in recent years, there has been a surge of research activity, mainly in the surface science field and with single-crystal catalysts, and a fairly clear picture of oxygen adsorption has emerged.

B. Single-Crystal Studies

Oxygen adsorption and desorption and nature of the adsorbed adlayer on silver single crystals have been studied by the modern techniques of surface science. More specifically, LEED and HREELS continue to be used in conjunction with temperature-programmed desorption (TPD), to study the adsorption of oxygen on individual low-indexed planes of silver single crystals.

Recent work utilizes AES,[68-72] LEED,[7,68,70,72-74,88,89] TPD,[68,69,72-75,95] ellipsometry,[78-80] UPS,[71,81,82] XPS,[71,72,76,83,84,88,89] ISS,[85] and HREELS[42,86,87,95] as well as work-function measurements;[73,74] experiments were conducted under ultrahigh vacuum (UHV) conditions. Much of the work has been carried out on the Ag(110) surface and is discussed in detail below, though adsorption of oxygen on Ag(111) surface has also been examined by several researchers. More recently, Campbell[72] has shown that the coverage of O_2(a) under reaction conditions on Ag(111) are very similar to Ag(110).

On Ag(111), Dweydari and Mee[126] found changes in the work function ϕ which indicates adsorption at room temperature at oxygen pressures below 10^{-6} torr. Engelhardt and Menzel[73] noted very small and irreproducible changes in ϕ, and ascribed these to a general inertness of the close-packed Ag(111) face and to some adsorption of oxygen at irregularities. In LEED and thermal desorption experiments, Rovida et al.[74] observed formation of a (4 $\times$ 4) pattern with clear evidence for oxygen adsorption in the Auger at temperatures above 370 K and pressures greater than 10^{-3} torr, and small and insignificant variations of ϕ at lower pressures. Albers et al.[79] observed similar LEED and desorption behavior following adsorption of oxygen on Ag(111) at 10^{-5} to 10^{-3} torr and the crystal temperature between room temperature and 520 K. These authors reported an extremely low sticking probability, 3×10^{-5} for well-annealed surfaces and 5×10^{-4} for damaged surfaces. Recently, Grant and Lambert[71] have established, by isotope-labeled thermal desorption, that oxygen adsorbs into a dioxygen species which coexist with adsorbed atomic oxygen. Based on XPS and Auger spectra, these authors concluded that O_2(a) resides on the Ag(111) surface whereas O(a) is located within the surface.

On the Ag(100) face, Engelhardt and Menzel[73] observed the adsorption of oxygen in a disordered structure. From ϕ measurements, two adsorption states were inferred on the surface in temperature-dependent equilibrium. The adsorbed oxygen caused no change in the clean surface LEED patterns, in agreement with the finding of Rovida et al.[74] However, Rovida observed facetting of (100) surface under higher oxygen pressure.

Extensive results have been obtained on the (110) face. Long-range interactions have been observed between oxygen atoms on this surface.[73,80] In particular, there are strong attractive interactions between oxygen atoms in the (001) direction, but strong repulsive interactions in the (110) direction. Superstructures, continuously changing (n $\times$ 1) LEED patterns with n ranging from 7 to 2, depending on the coverage, have been observed. Furthermore, the kinetics of oxygen atoms recombination are quite complex due to the lateral interactions, showing an apparent variation in reaction order.

In the following pages these observations are discussed as they specifically apply to the Ag(110) surface. Some selected experimentations are also described.

C. Atomically Adsorbed Oxygen Only

At temperatures above 150 K, oxygen adsorption is dissociative.[76] The oxygen-sticking probability diminishes continuously with increasing coverage. Albers et al.[80] found an initial sticking coefficient of 5.9×10^{-4} at 294 K, which decreased to 4.1×10^{-4} at 475 K.

Engelhardt and Menzel,[73] and Bowker, Barteau, and Madix[75] found an initial sticking coefficient of 3×10^{-3} at 300 K, which also decreased with increasing temperature of the crystal.

The adsorbed oxygen atoms have been shown previously to form chains along the (001) direction of the surface which give rise to (n × 1) LEED patterns.[73] An earlier report[68] had also indicated that LEED features characteristic of a faint (1 × 2) pattern were present following oxygen adsorption on the Ag(110) surface. The result of Barteau's study[82] clearly showed that adsorbed oxygen atoms produced only (n × 1) LEED patterns. Examples of the (4 × 1), (3 × 1), and (2 × 1) LEED patterns are shown in Figure 5. These surface oxygen structures were due to the formation of chains of oxygen atoms along the (001) direction on the surface which interact repulsively along the (110) direction.[82]

Several other LEED features for oxygen adsorption on Ag(110) have been reported.[7,68,90] Small amounts of CO contamination have been shown to alter the work function behavior of the oxygen adlayer, and splitting of the half-order spots in the (2 × 1) pattern and observation of a faint (1 × 2) pattern have been attributed to CO adsorption.[73,90,91] Barteau and Madix[92] have subsequently shown that the additional features in the LEED patterns, p (1 × 2) symmetry, are due to CO_3 formation. Figure 6 shows their LEED patterns observed for reaction of CO_2 with approximately 1/3 monolayer of oxygen atoms. They concluded that reaction of adsorbed oxygen with CO_2 at 300 K results in the formation of (1 × 2) domains of adsorbed carbonate, and compression of the oxygen chains into a (2 × 1) structure. This stoichiometry is consistent with other measurements.[88,95]

The thermal desorption of oxygen from the Ag(110) surface has also been examined in the literature. Although studies of oxygen atom recombination and silver powders[55] have shown second-order kinetics, the strong lateral interactions which give rise to the formation of oxygen atom chains or the Ag(110) surface strongly perturb the recombination kinetics.[7,73,75,77,82] Thermal desorption spectra for oxygen atom recombination on Ag(110) are shown in Figures 7 and 8.[7] For low initial oxygen coverages (Figure 7), the oxygen atoms are too dispersed for chain formation to occur, and the recombination does exhibit the second-order kinetics. However, for oxygen coverages greater than 30 L (Figure 8), the desorption peaks exhibit pseudo first-order behavior. One Langmuir (L) is equal to an exposure at 10^{-6} torr for 1 sec. The deviations from second-order kinetics are shown most clearly by the dependence of the peak temperature on coverage (Figure 9[7]); the peak temperature passes through a minimum at coverages of the order of 10^{13} atoms/cm^2. Other studies[73,77] have also shown that the thermal desorption peak for oxygen shifts to a higher temperature as the initial coverage is increased. For a true first-order process the peak maximum should remain in the same position. The interpretations of these studies are, however, questionable. For example, Rovida[37] interpreted the desorption in terms of second-order rate law, although over the range of oxygen coverages examined such conclusion was consistent with neither the shape of the desorption peak nor the dependence of the peak temperature upon the initial coverage.[127] To date, a complete description of the desorption model which incorporates all of these considerations is still lacking.

In a novel experiment, Campbell and Paffett[88] populated a state of atomically adsorbed oxygen by dosing 50 torr O_2 and 485 K in a high-pressure cell, and then rapidly transferred back the crystal into UHV. A coverage of $\theta_0 = 0.67$ atomically adsorbed oxygen was achieved, which gave a new c (6 × 2) overlayer structure in LEED and a thermal desorption peak at 565 K, in addition to the standard O_2-TDS state at about 600 K. Upon removing the 565 K peak by flashing to 575 K, the p (2 × 1) LEED pattern reappeared. The oxygen in the c (6 × 2) pattern was atomically adsorbed which was further verified by XPS.

D. Atomically and Molecularly Adsorbed Oxygen

The very recent work in this area has clarified the nature of the oxygen adsorbed on single-

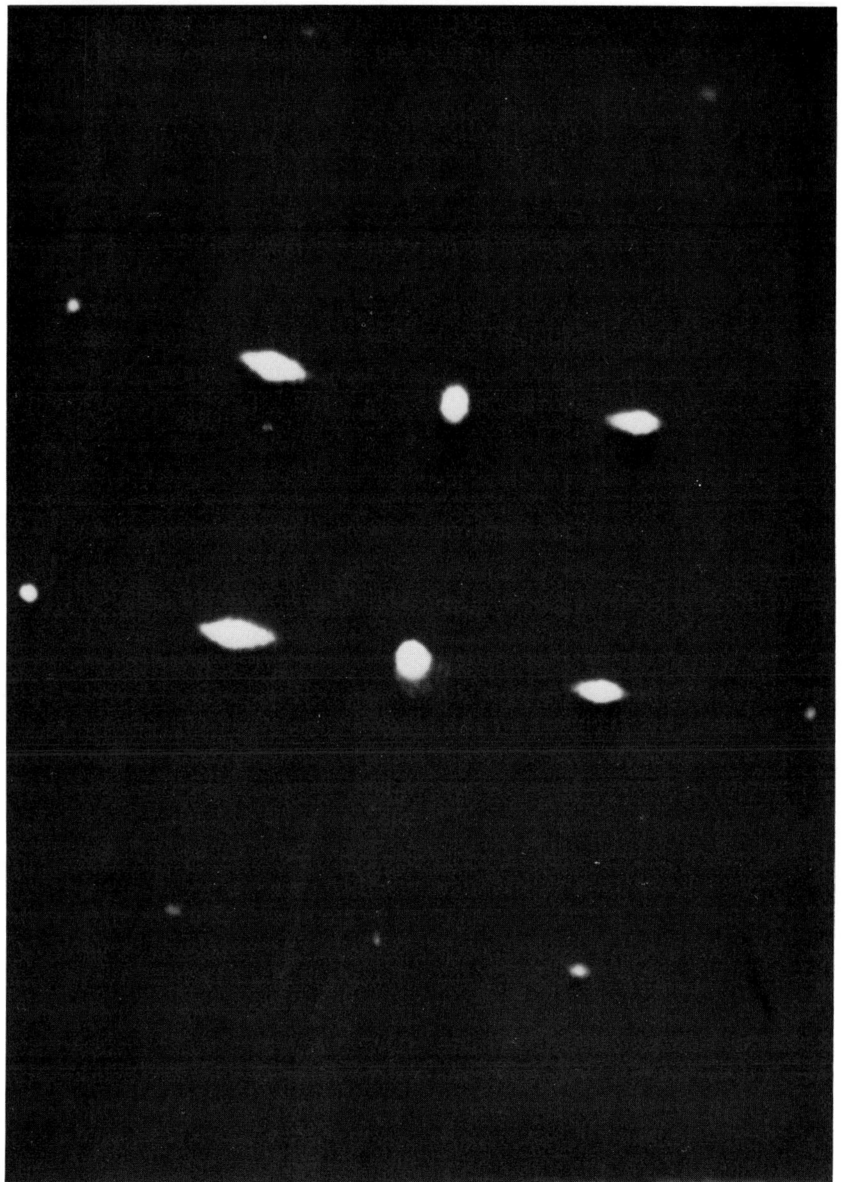

A

FIGURE 5. (n × 1) LEED patterns following adsorption of oxygen on the Ag(110) surface at 300K: (A) (2 × 1), exposure = 6000 L, beam energy = 51 eV; (B) (3 × 1), exposure = 600 L, beam energy = 30 eV; (C) (4 × 1), exposure = 240 L, beam energy = 48 eV. (From Barteau, M. A. and Madix, R. J., in *The Chemical Physics of Solid Surfaces and Heterogeneous Catalysis*, King, D. A. and Woodruff, D. P., Eds., Elsevier Science Publishers, Amsterdam, 1982, chap. 4. With permission).

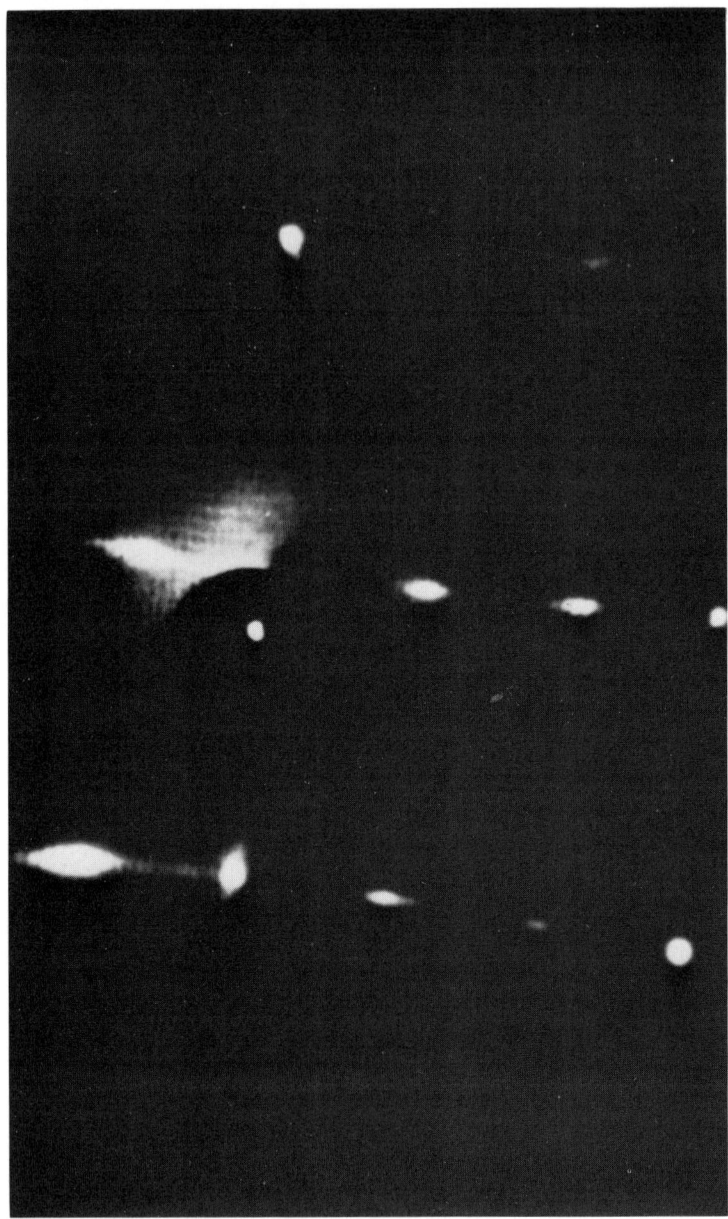

FIGURE 5B

crystal silver surfaces. Thus, Barteau and Madix[76] have shown, by isotopic exchange thermal desorption, that while the adsorption of oxygen on Ag(110) above 150 K was dissociative and resulted in a desorption peak at 590 K, when the sample was cooled to 150 K or below before exposure to oxygen, a second, low-temperature oxygen molecular state was observed to desorb at 190 K (desorption activation energy, E_d = 45 kJ/mol). Figure 10 shows thermal desorption results following exposure of the surface at 133 K to 60 L of $^{16}O_2$ followed by 540 L of $^{18}O_2$. The ratio of ^{16}O to ^{18}O at 590 K was 2.2, while that of $^{16}O_2$ to $^{18}O_2$ was 0.56, indicating that at higher exposures, oxygen adsorption into the molecular state was favored. No ordered LEED structures other than the p (1×1) were reported following oxygen adsorption at 130 K.

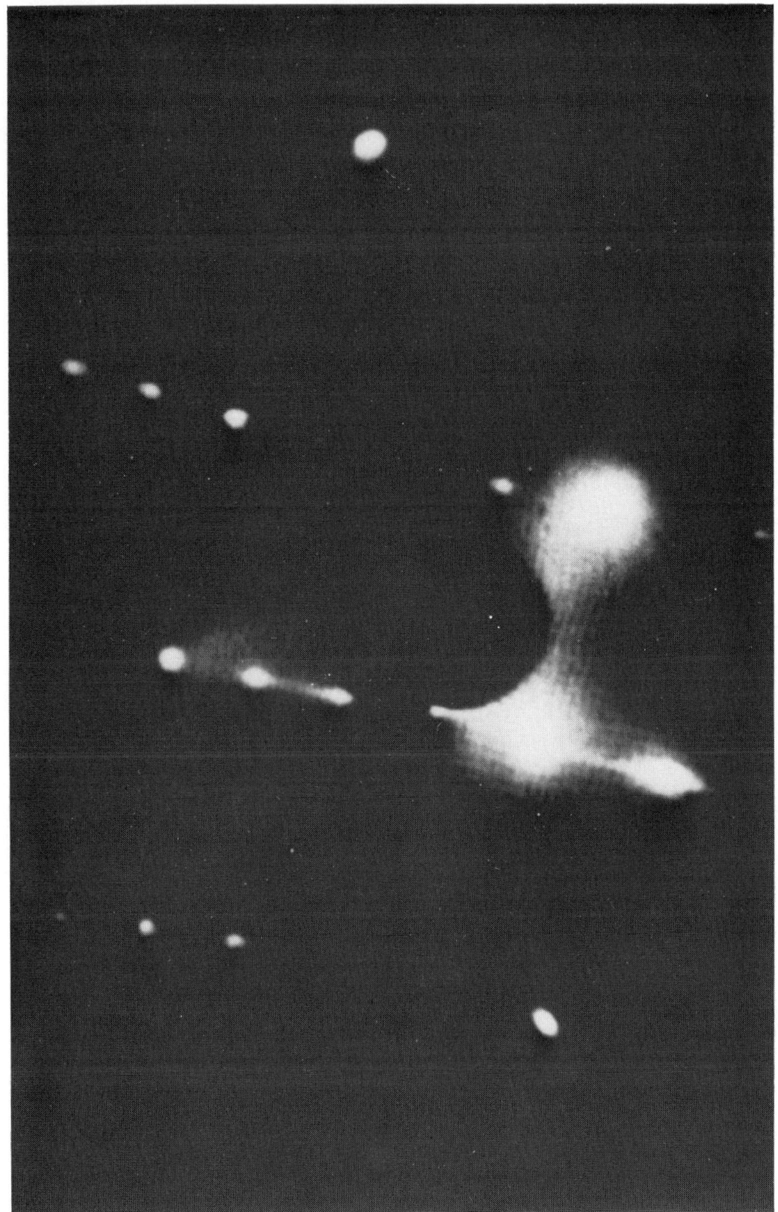

FIGURE 5C

The initial sticking probability for oxygen at 126 K was reported to be 8×10^{-3} based on the total oxygen coverage, whereas sticking probability at 300 K was 3×10^{-3}. This increase was not simply due to the additional binding state for oxygen.[7,76] These authors have shown (Figure 11) that the coverage of the high-temperature state alone was greater at low exposures following adsorption at 126 K than for adsorption at 300 K at low temperature as opposed to one at higher temperatures.

Several authors[86,87] have also examined the adsorption of molecular oxygen on Ag(110) at temperatures of 100 K. Sexton and Madix[87] observed oxygen adsorption at 100 K to be almost entirely molecular, although a small contribution to the HREELS spectra from O(a)

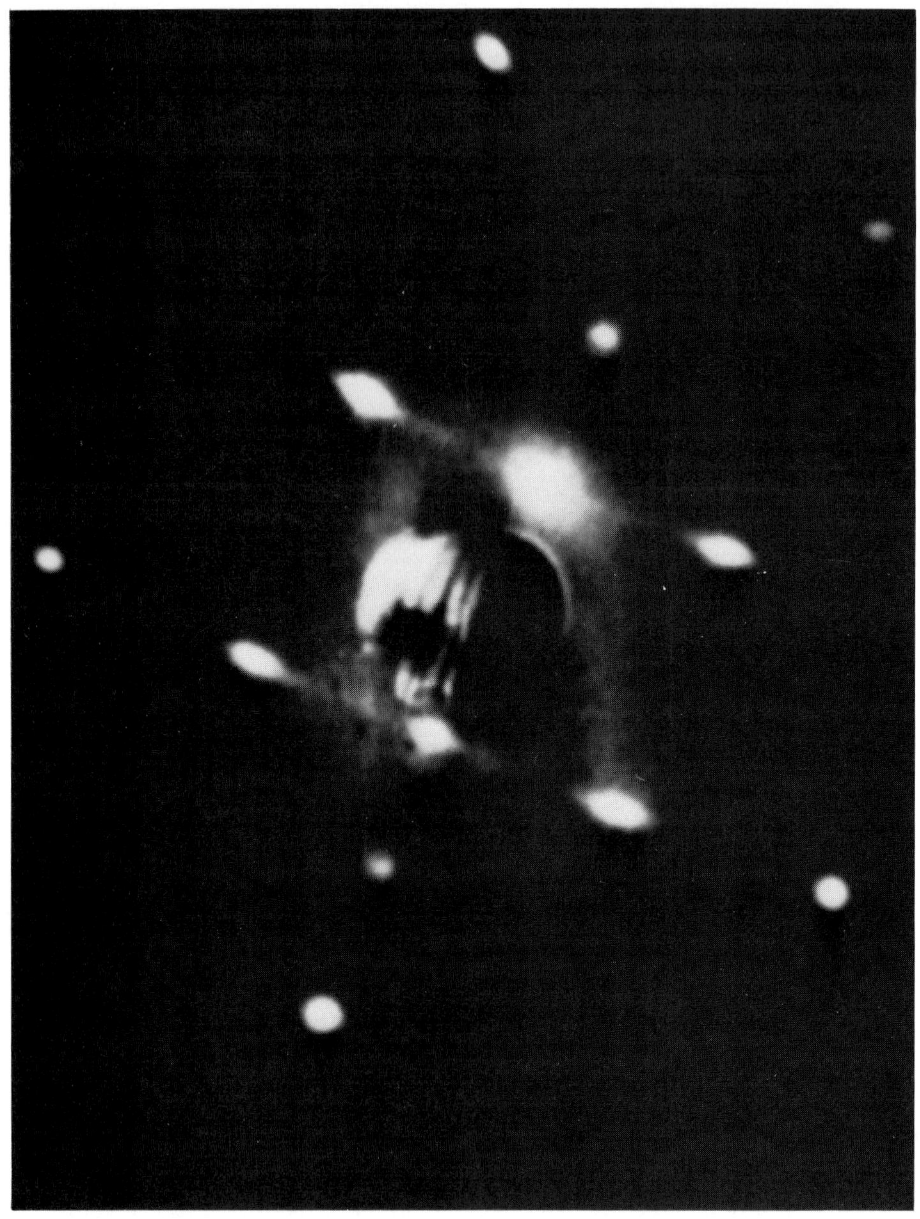

A

FIGURE 6. LEED patterns observed following exposure of the Ag(110) surface to 600 L O_2 followed by CO_2 at 300 K: (A) CO_2 exposure = 240 L, beam energy = 45 eV; (B) CO_2 exposure = 600 L, beam energy = 35 eV; (C) p(3 × 1) oxygen LEED pattern following heating of the surface in (B) to 520 K to decompose CO_3 species, beam energy = 46 eV. (From Barteau, M. A. and Madix, R. J., *J. Chem. Phys.*, 74, 4144, 1981. With permission.)

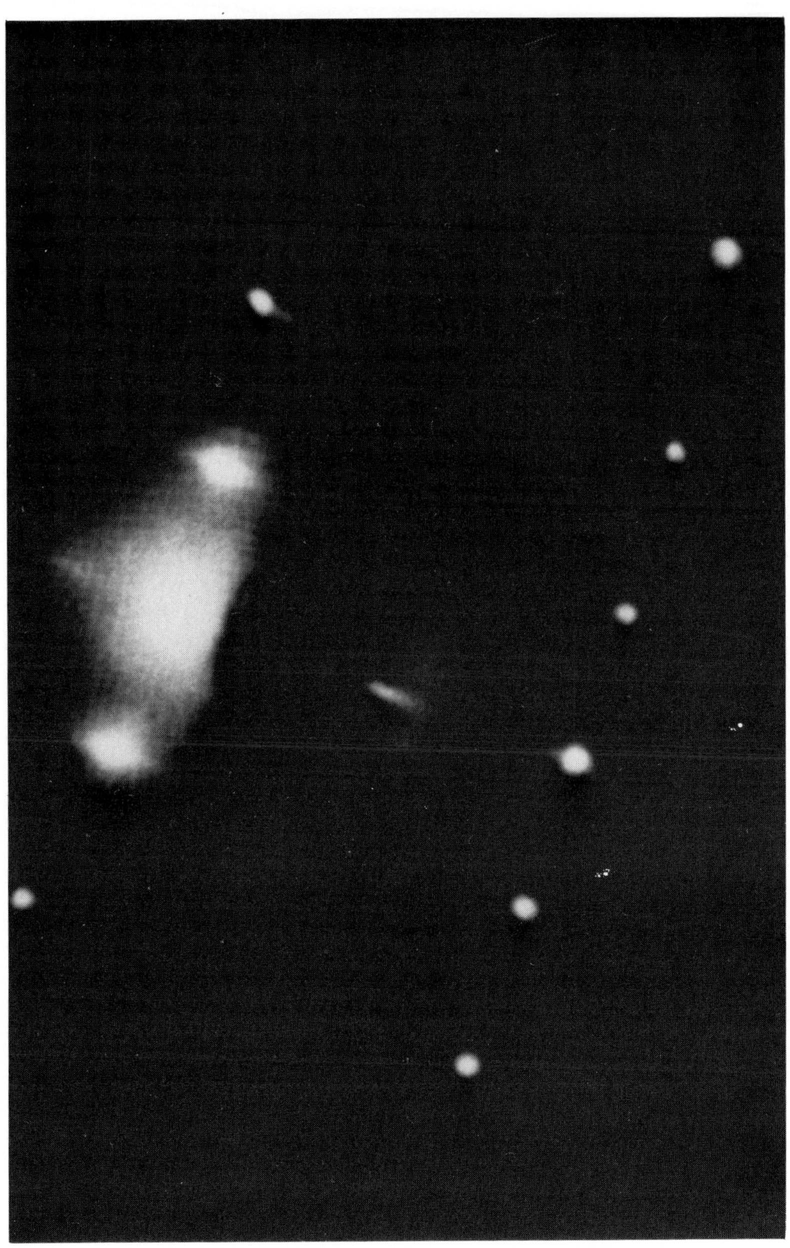

FIGURE 6B

could not be eliminated. Backx et al.[86] have identified three different types of adsorbed oxygen species on the Ag(110) face in the temperature range 110 to 770 K. The relevant HREELS spectra is presented in Figure 12. At 110 K, oxygen was adsorbed as a diatomic species. Two losses were detected: at 30 and 78 meV. No loss due to atomic oxygen was reported at 39 meV. Heating the crystal to above 170 K, the temperature at which diatomic oxygen is desorbed, changed the spectrum drastically into a single loss at 39 meV. Further heating of the crystal to 310 K resulted in an increase of the intensity of the 39 meV peak. Above 420 K, adsorbed atomic oxygen migrates from the surface to subsurface.

FIGURE 6C

In order to reconcile with the high-temperature results of the three studies, all that is required is the assumption that the presence of oxygen atoms on the surface inhibits the dissociation of oxygen molecules. Thus, the substantial oxygen atom coverages created upon adsorption at 125 K prevent further dissociation of the molecular oxygen during thermal desorption, whereas adsorption at 100 K produces an insufficient coverage of oxygen atoms to prevent dissociation upon heating.[82]

Other surface techniques have also provided information on the adsorption of oxygen on Ag(110) as well as on silver foil. Heiland et al.[85] studied the (2 × 1) structure using ion-scattering spectroscopy. From the strong azimuthal anisotropy of the ISS signal, they con-

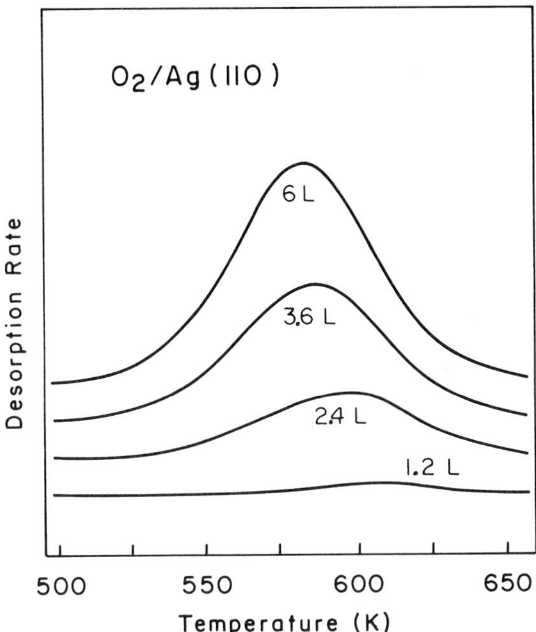

FIGURE 7. Thermal desorption spectra for O_2 adsorption on Ag(110) at 300 K; low-oxygen exposures. (From Barteau, M. A. and Madix, R. J., in *The Chemical Physics of Solid Surfaces and Heterogeneous Catalysis*, King, D. A. and Woodruff, D. P., Eds., Elsevier Science Publishers, Amsterdam, 1982, chap. 4. With permission.)

cluded that the oxygen atoms are positioned between two Ag atoms in the (110) surface channels. Published data on the UPS of the oxygen-silver system show significant variations depending on the exact nature of the surface and the conditions of adsorption.[81,82,94-96] Quantitative explanation of the UPS features have not been proven successful.

XPS results[82,88,97] have shown the existence of three oxygen species on silver. They have been assigned to oxygen atoms, subsurface oxygen, and molecularly adsorbed oxygen. The binding energies are listed in Table 2. O(1s) binding energies for all three species are less than 531 eV, although binding energies (O[s]) such as 533 and 531.8 to the species O_2(a) and O(a), respectively, have also been assigned.[84] It appears that some of the XPS data have to be reinterpreted where O(1s) peak assignments were made without the help of substantial supporting techniques.

The activation energy reported for desorption is not particularly sensitive to kinetic model assumed. Values reported for Ag(110) are in good agreement with each other and at the upper end of the range reported for desorption of oxygen from silver powders (Table 3).

Based on the preceding observations,[75,76,86] Bowker[93] suggested an elaborate scheme for the oxygen-Ag(110) interaction. The entire sequence of the potential energy scheme is shown in Figure 13. In this scheme, a third state of oxygen (O_L) is shown to exist. The adsorption process, above the desorption peak for the molecular state (>200 K), is shown below:

$$O_2(g) \rightarrow O_2(a) \tag{7}$$

$$O_2(a) \rightarrow 2O(a) \tag{8}$$

$$O_2 \rightarrow O_L \tag{9}$$

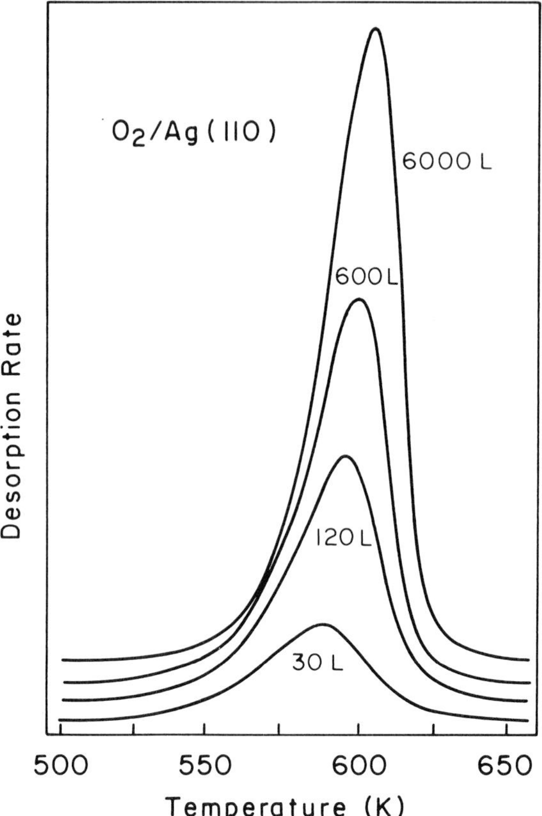

FIGURE 8. Thermal desorption spectra for O_2 adsorption on Ag(110) at 300 K; high oxygen exposures. (From Barteau, M. A. and Madix, R. J., in *The Chemical Physics of Solid Surfaces and Heterogeneous Catalysis,* King, D. A. and Woodruff, D. P., Eds., Elsevier Science Publishers, Amsterdam, 1982, chap. 4. With permission.)

Although the third state of oxygen is not seen directly, it has been included to account for some of the AES[68] and desorption[86] observations. For example, Rovida and Pratesi[68] had observed an Auger signal which persisted after desorption of the atomic surface state.

These results, combined with the preceding discussion, strongly suggest that molecular oxygen, O_2(a), desorbs at around 190 K, and atomically adsorbed oxygen, O(a), associatively desorbs at around 600 K. A third state of oxygen, probably subsurface form, also appears to exist on silver surfaces. The question now arises that what oxidation states do these surface atoms achieve under ethene epoxidation reaction conditions, and which are preferred. These discussions may also address the suspected role of subsurface oxygen. Thus, the concluding section focuses on the reaction mechanism.

VI. REACTION MECHANISM

A. Unpromoted Silver

The exact mechanism of Reaction 2 has remained unknown, but several different ones based on well-defined silver surfaces have been recently proposed.[43,44,48,70,94,97-99] There is general agreement that the formation and decomposition of surface carbonate on silver surface by the interaction of CO_2 gas with atomically adsorbed oxygen proceeds with the stoichi-

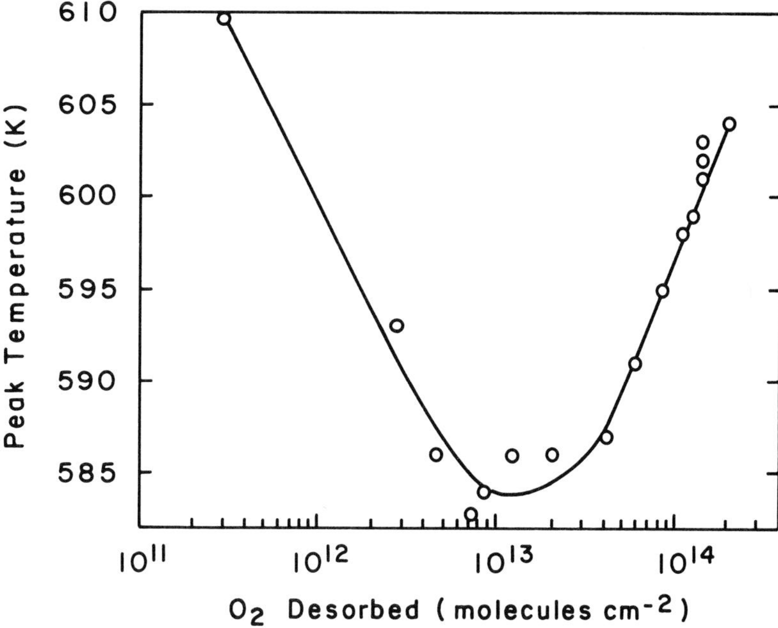

FIGURE 9. Dependence of the peak temperature for O_2 desorption from Ag(110) upon the oxygen coverage. (From Barteau, M. A. and Madix, R. J., in *The Chemical Physics of Solid Surfaces and Heterogeneous Catalysis,* King, D. A. and Woodruff, D. P., Eds., Elsevier Science Publishers, Amsterdam, 1982, chap. 4. With permission.)

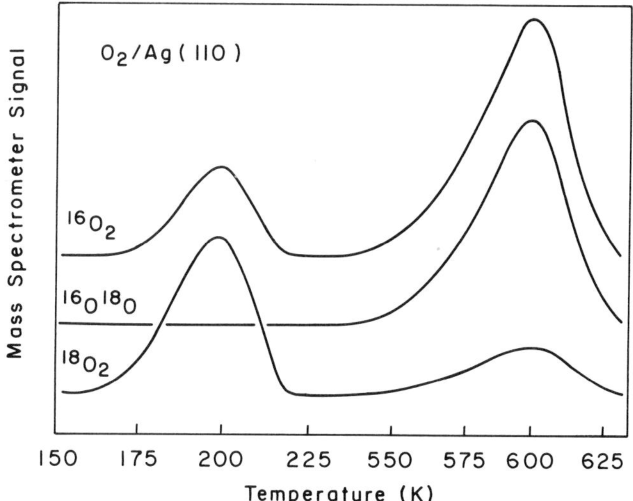

FIGURE 10. Thermal desorption of oxygen from Ag(110) showing isotopic exchange in the atomic state (590 K) and no exchange for the molecular state (200 K). (From Barteau, M. A. and Madix, R. J., *Surf. Sci.,* 97, 1, 1980. With permission.)

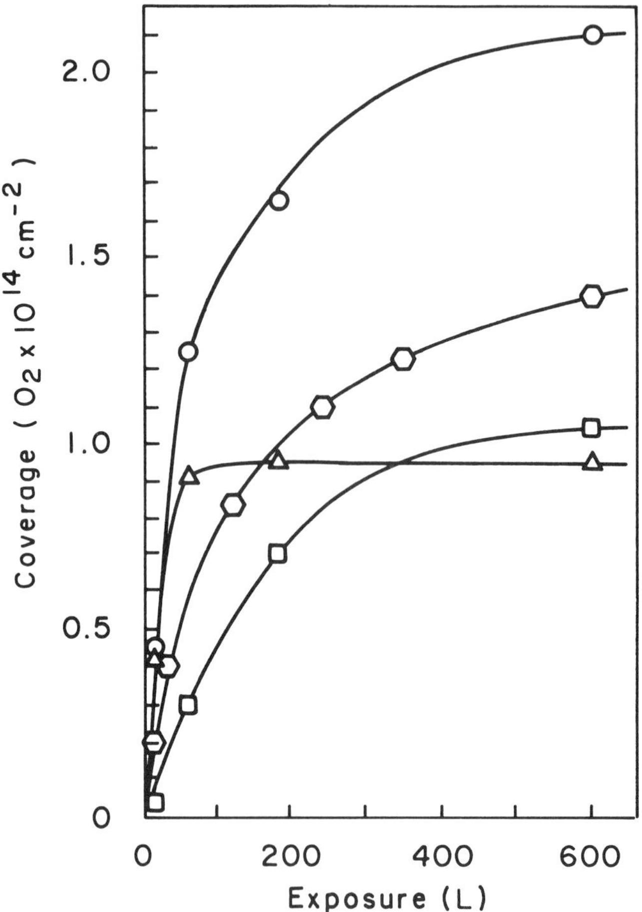

FIGURE 11. Coverage vs. exposure profiles for oxygen adsorption on Ag(110). Ta = 126 K: □ O_2(190 K), △ O_2(590 K), ○ Total oxygen coverage. Ta = 300 K: ⬡. Total oxygen coverage (590 K). (From Barteau, M. A. and Madix, R. J., *Surf. Sci.*, 97, 101, 1980. With permission.)

ometry CO_2 + O(a) = CO_3(a). The presence of oxygen atoms on silver surface alters the adsorption behavior.

Fundamental studies of the interaction between ethene and oxygen on silver have been performed only recently. For the mechanism of Reaction 2, basically three schools of thought have emerged from these studies. Shell researchers,[47,94] based on single-crystal studies, have concluded that atomically adsorbed oxygen is active in the formation of ethene oxide. More recently, they[48] have shown that, when deposited to high coverage ($0.5<\theta_o<1$) on silver powder, preadsorbed oxygen adatoms could be transiently converted with high selectivity to ethene epoxide using only gas phase ethene and no O_2. Thus, Reactions 3 and 4 appear to be an inadequate description of the selective oxidation route.

The second picture emerges from the work of Campbell et al.[72,88,89,99] on single-crystal studies, from which it is concluded that there is no relation between concentration of atomic oxygen which is observed to desorb at 600 K in TPD experiments, and epoxidation rate. They showed that their results are consistent with a mechanism whereby molecularly adsorbed oxygen and ethene combine to form an intermediate in the rate-determining step. This intermediate then rapidly branches into ethene epoxide or CO_2 pathways. The mechanism is shown below:

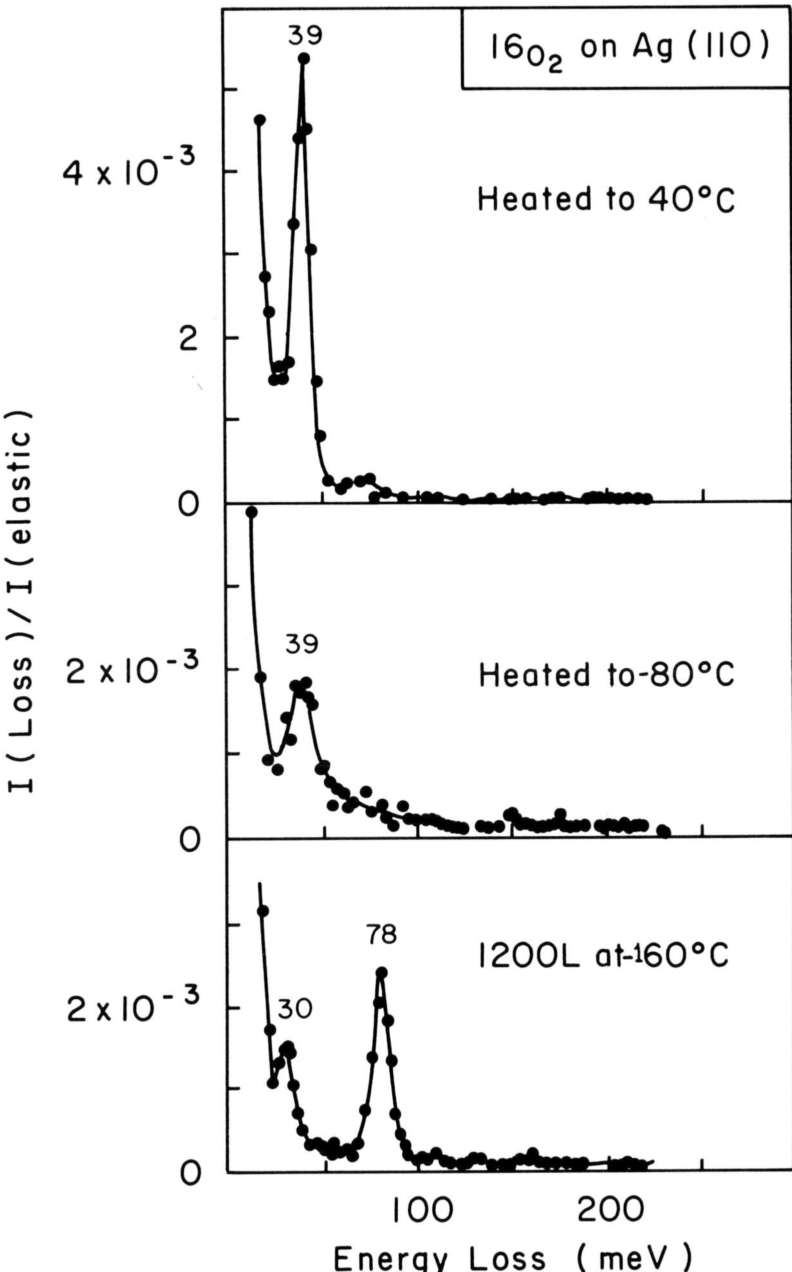

FIGURE 12. HREELS spectra of Ag(110) showing the conversion from the molecular state present at low temperatures (losses at 30 and 78 meV) to the atomic state present at room temperature (one loss at 39 meV). (From Backx, C., De Groot, C. P. M., and Biloen, P., *Surf. Sci.*, 104, 300, 1981. With permission.)

Table 2
O (1s) BINDING ENERGIES FOR OXYGEN SPECIES ON SILVER

	O(1s) Binding Energy (eV)	
Species	**Ag(110)**	**Ag foil[97]**
O(a)	528.3[128]	528.3
	528.1[88]	
O(subsurface)	531.2[128]	530.3
	528.5[88]	
O_2(a)	529.9[128]	532.5
	529.3[88]	

Table 3
ACTIVATION ENERGIES FOR DESORPTION

Silver	Value (kJ/mol)	Ref.
(110)	169	77
(110)	172	73
(110)	173	75
(110)	167	88
(111)	146—168	74
(111)	130	71
Powder	133—175	55

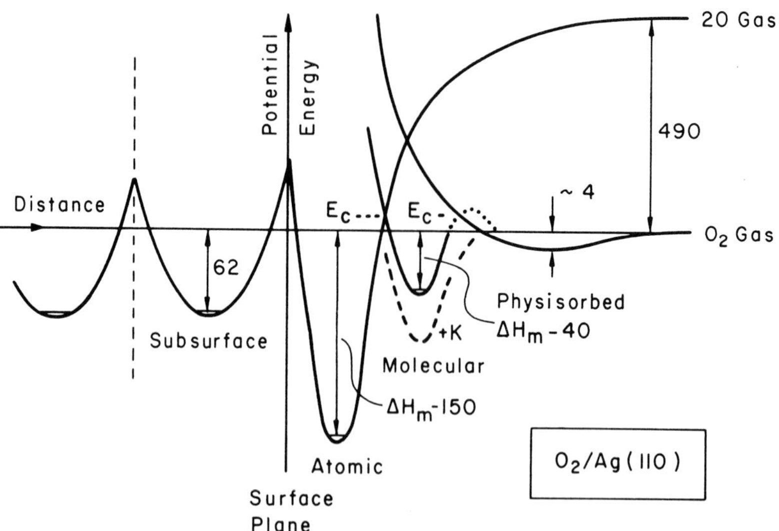

FIGURE 13. A simplified one-dimensional potential energy scheme showing the possibility of four types of oxygen species adsorbed on a silver(110) surface. (Reprinted with permission from Bowker, M., *Vacuum*, 33, 669, Copyright 1983, Pergamon Press, Ltd.)

$$C_2H_4 \overset{\text{Fast}}{\rightleftharpoons} C_2H_4(a) \tag{10}$$

$$O_2 \overset{\text{Fast}}{\rightleftharpoons} O_2(a) \rightleftharpoons 2O(a) \tag{11}$$

$$C_2H_4(a) + O_2(a) \xrightarrow{\text{Slow}} I(a) \tag{12}$$

$$I(a) \nearrow C_2H_4O + O(a) \tag{13a}$$

$$I(a) \searrow^{\text{Fast}} \text{Fragments} \xrightarrow{O(a)} CO_2 + H_2O \tag{13b}$$

The selectivity is determined by the branching ratio of fast steps leading to the decomposition of the adsorbed intermediate [I(a)].

The last observation that is described here is that of Grant and Lambert.[71,98] Based on a well-characterized single surface at pressures up to 50 torr, they proposed that chemisorbed atomic oxygen reacts with adsorbed ethene to yield both ethene oxide and combustion products. Chemisorbed dioxygen, though present, appears to play no direct role in either of these reactions; the presence of subsurface oxygen is necessary for selective oxidation but not for total oxidation.

Ethene oxide can be oxidized to CO_2 and H_2O over supported silver catalysts, and this process may also have an effect in the overall selectivity of the epoxidation. Although weak adsorption has been reported for EO on silver and silver preexposed to oxygen, it remains unclear whether this occurs mainly on the Ag surface of on the support.[94] Recently, Grant and Lambert[113] have shown that EO isomerises on oxygen-containing Ag(111) surfaces which are active for ethene epoxidation.

We thus have an interesting but precise picture of the epoxidation process. However, before attempting to generalize a mechanism, we will review the recent literature concerning the role of promoters, especially that of chlorine and the alkali metals. The former plays a major role in EO synthesis since this has effects on the adsorbed oxygen species.

B. Moderation by Chlorine

As presented earlier in the text, trace quantities of chlorinated hydrocarbons are added to the reactant feed which considerably improves the selectivity of the reaction. The role of these chlorine promoters remains unclear.

Ostrovskii et al.[13] studied the effect of Cl additive to silver catalysts at concentrations in the range of 10^{-5} to 10^{-1} at.% (Figures 14a and b). Included in this figure (14c) are also the effects of S, Se, and Te additives to silver catalysts. At low concentration, Cl increased the catalytic activity; the catalyst is poisoned at higher concentrations. The changes in catalytic activity and selectivity brought about by the modification were explained as resulting from changes in the bond energies of the adsorbed oxygen. The role played by S, Se, and Te additives are controversial.

The effect of chlorine in the Sachtler-modified Worbs mechanism (Figure 3)[12,35] was thought to be to block sites from the adsorbed diatomic precursor to adsorbed monoatomic oxygen. The dissociation would therefore require four adjacent silver atoms; one quarter of a monolayer is enough to inhibit this process completely.

This interpretation conflicts with that of Rovida et al.[63] On the Ag(110) face, the presence of preadsorbed chlorine reduces the amount of adsorbed oxygen and, when the chlorine coverage is greater than about 0.5, oxygen adsorption is completely inhibited. Their interpretation of the LEED spectra suggests that chlorine and oxygen coexist on the surface mainly in separate domains, each with its own individual structure. They further concluded that the presence of preabsorbed chlorine substantially weakens the silver-oxygen bond.

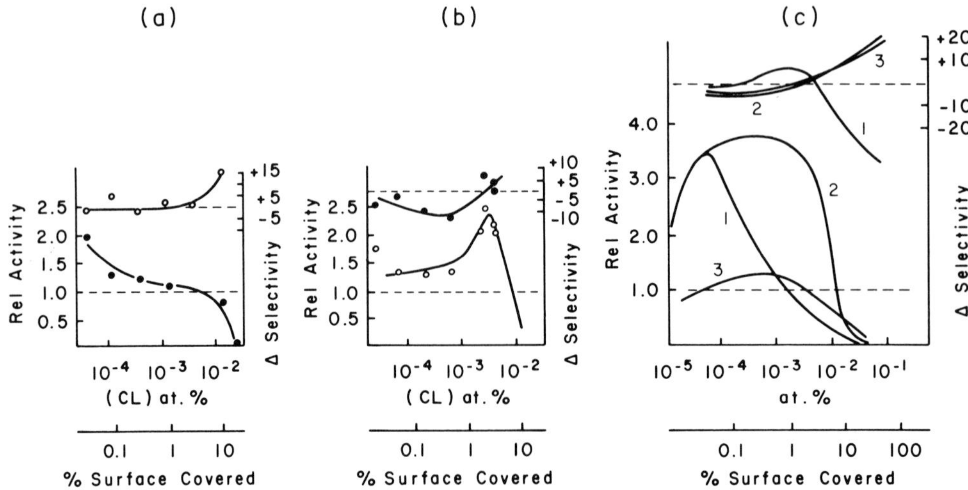

FIGURE 14. Dependence of catalyst activity and selectivity on (a) chlorine addition at 488 K (b) chlorine addition at 573 K; (c) on additive concentration: (1) sulfur (2) selenium (3) tellurium.[13]

In the past 5 years, several studies for the Cl_2-Ag system with single-crystal catalysts have been reported. The recent work utilizes LEED,[100-107,126] AES,[102-107] and XPS,[105-107,126] as well as TPD.[103,104] On the Ag(100) plane, Kitson and Lambert[100] found that the chlorine adsorbed dissociatively, adsorption terminated at a monolayer, and the chlorine was desorbed in the form of AgCl at approximately 850 K. Some work on Ag(111) was also pursued by the Lambert group without complete results.[101] They nevertheless suggested the adsorbed chlorine to desorb as atomic Cl. LEED studies reported a (10 × 10) arrangement[101] and a pattern interpreted in terms of the formation of a AgCl(111) superlattice.[102]

Chlorine adsorbs dissociatively with an initial sticking probability of 0.4 and desorbs as a compound AgCl.[103,104] The characteristics of chlorine adsorption on Ag(110) were reported to be similar to those for Ag(111). Their thermal desorption spectra, after chlorine adsorption on the Ag(111) surface at 300 K for moderate exposures (up to 120 L), are shown in Figure 15.[103] Desorption from a surface with submonolayer quantities of chlorine adsorbed indicated a strongly bound species which desorbed at approximately 780 K, exclusively as AgCl. The lower temperature peak (670 K) corresponds to the presence of multilayers. The corresponding results of temperature-programmed Auger measurements in the monolayer and multilayer regimes showed a complexity, as illustrated in Figure 16. These authors, however, concluded that under synthesis condition, it is likely that only surface chlorine is present on the catalyst, though chlorine can dissolve into the surface lattice only after the formation of the surface layer.[93] The desorption does not occur at a significant rate until around 650 K — roughly 130 K above the reaction temperature.

Campbell's group[105-107] has recently modeled the role of chlorine promoters by depositing chlorine adatoms onto the Ag(110) and (111) surfaces in UHV, employing XPS, LEED, and AES, and monitoring the resultant effects on the medium-pressure (approximately 1 atm) kinetics of the epoxidation reaction. The Ag(111) surface was found to behave differently than Ag(110) in selectivity with respect to chlorine coverage but, nevertheless, displayed similar catalytic behavior for epoxidation (Figure 17). Chlorine adatoms have been postulated to have electronic effects on the heats of adsorption of ethene and molecular O_2, which alter the steady-state coverages of these species under reaction conditions. This changes the rate of slow step (Reaction 12 above), which is, however, unrelated to the selectivity. These authors have concluded that the selectivity is controlled by the chlorine coverage via an ensemble effect on the rapid branching ratio (Reaction 13 above). Fragmentation of the

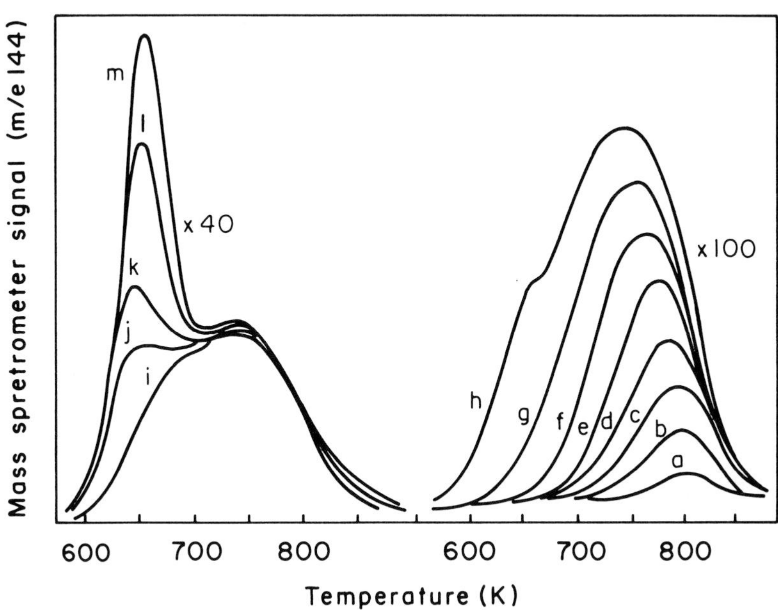

FIGURE 15. Mass spectrometric thermal desorption spectra of silver chloride after chlorine adsorption on the Ag(111) surface at 300 K. The exposures to chlorine (in L) are (a) 0.4, (b) 0.8, (c) 1.0, (d) 1.2, (e) 1.6, (f) 2.4, (g) 3.0, (h) 4.8, (i) 9.6, (j) 48, (k) 84, (l) 102, and (m) 120. (From Bowker, M. and Waugh, K. C., *Surf. Sci.*, 134, 639, 1983. With permission.)

intermediate (13b) required larger ensembles of chlorine-free Ag sites than does the epoxidation branch (13a). This model is also in contrast to Sachtler's mechanism,[35] which attributed the enhancement of selectivity entirely to the effect of chlorine on the dissociation rate of O_2. Such a model thus becomes questionable since the latter effect occurs at low chlorine coverage, while the former effect is only at high coverage.

C. Alkali Metal Promoters

In industrial processes, alkali metal salts are added in catalyst preparation in order to increase the selectivity. A large number of patents have been issued on the incorporation of alkali metals in the silver catalyst and all seem to produce promotion effects, as discussed earlier in the text. Patent literature generally describes the addition of alkalis from a water solution, followed by heating and drying in air. Certain alkali metal-promoted catalysts have been found to enhance the selectivity to the 78% region,[116] and some of them even higher than the maximum theoretically achievable by a reaction following Worbs-Sachtler mechanism (Figure 3). The exact function of these promoters are far from clear.

Numerous studies have been conducted over the past 10 years on single crystals.[108-115] By far, the most comprehensive work is that of Lambert, who has examined alkali metal-dosed single-crystal surfaces.[108-112] It is clear from their studies that the adsorption properties of oxygen varies with coverage of the promoter.[108-113] Spencer and Lambert[115] have shown that the rubidium diffuses with a very low activation energy into the immediate subsurface region, while simultaneously enhancing the sticking probability for oxygen on this face by up to seven orders of magnitude. In fact, Lambert et al.[108-112] have observed the mutual enhancement of surface-to-bulk transport during coadsorption of all the alkali metals and oxygen on single-crystal surfaces. Thus, Figure 18 reveals the effect of bulk contamination

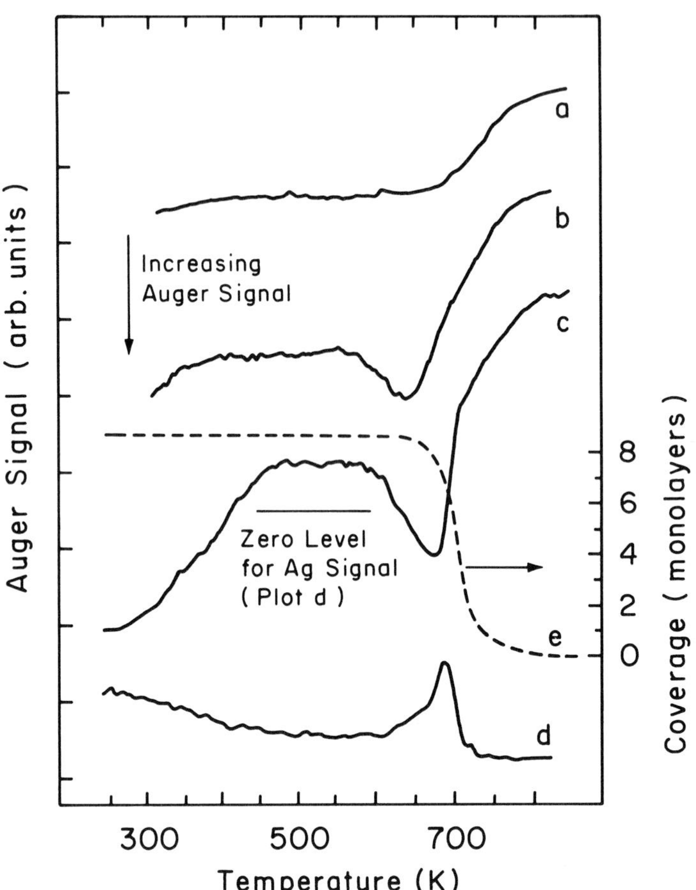

FIGURE 16. Temperature-programmed Auger spectra for chlorine layers adsorbed on Ag(111). (a) coverage of 0.8 monolayers adsorbed at 300 K, (b) coverage of 4.5 monolayers at 300 K, and (c) for a coverage of 9 monolayers at 245 K; (d) response of the Ag peak at 352 eV corresponding with (c); (e) coverage profile as a function of temperature for the experiment of curves c and d. (Reprinted with permission from Bowker, M., *Vacuum,* 33, 639, Copyright 1983, Pergamon Press, Ltd.)

by potassium on the oxygen desorption on Ag(100).[111] This suggests that subsurface oxygen may be present and perhaps plays an important mechanistic role in the epoxidation of ethene.

More recently, Campbell[114] has studied the role of cesium promoters by using adsorbed metallic cesium on the surface of Ag(111). They have shown that under reaction conditions, the Cs converts to a surface cesium oxide, C_sO_3 with a specific LEED pattern. In fact, the alkali promoters are almost always oxidized (in Lambert's work as well).

In industrial practice, chlorine and alkali metal promoters (particularly Cs) coexist, therefore, it is possible that they may be bonded together on the catalyst surface. Such a synergistic effect might arise from formation of coadsorbed structure or compound of these species, as has been proposed for KCl/Ag(100).[112] Also, cesium additives in real catalysts are deposited from their salt solutions and they, therefore, may have another structure to that which the above studies attain via vapor deposition.

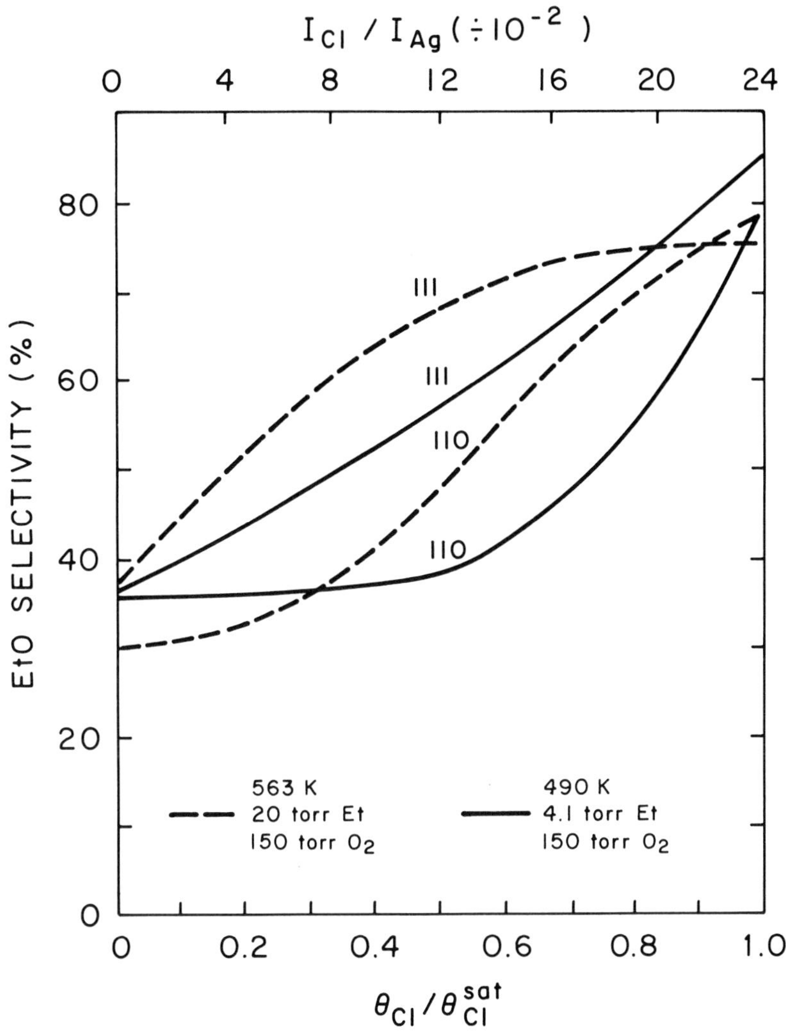

FIGURE 17. Selectivity variations with chlorine coverage for Ag(110) and Ag(111) faces. (From Campbell, C. T., *J. Catal.*, 99, 28, 1986. With permission.)

D. Summary and Mechanistic Questions

The foregoing background information, complemented by recent results of work on adsorption of oxygen on well-defined surfaces, allows us to formulate a picture of the epoxidation of ethene. We will assume that all the faces of silver contribute to the formation of ethene oxide.

The triangular reaction network in Equation 2 is the simplest set of reaction pathways which adequately describe the silver-catalyzed oxidation of ethene. As noted above, this network can be thought of as a series-parallel scheme: the epoxidation and combustion of ethene are parallel processes; the combustion of ethene oxide occurs in series with its formation. Much of the fundamental work aimed at resolving the mechanisms of these reactions has focused on the parallel processes, e.g., the surface oxygen species responsible for epoxidation and combustion of ethene. A substantial fraction of the efforts directed toward the improvement of catalyst performance have involved manipulation of the series processes; the choices of support composition, surface area, and metal loading may be governed by the need to minimize combustion of the product ethene oxide. Promoters may

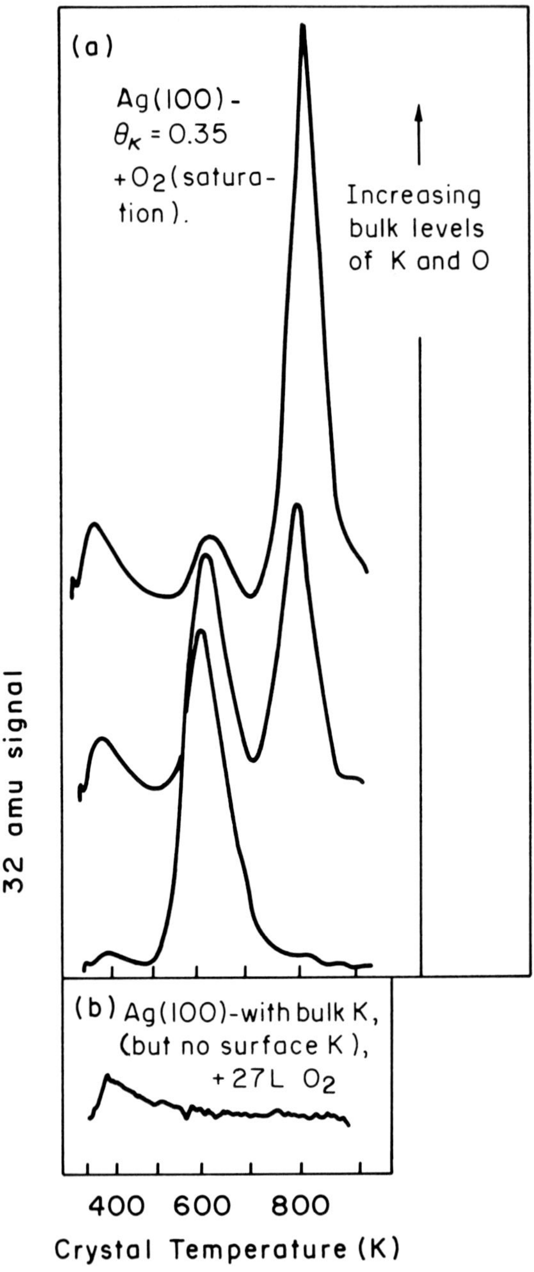

FIGURE 18. The effect of subsurface potassium on the desorption of oxygen from Ag(100). (From Kitson, M. and Lambert, R., *Surf. Sci.*, 109, 60, 1981. With permission.)

influence this reaction as well as the other two reaction components of the triangular network. This dichotomy of emphasis is perhaps responsible for the ironies surrounding this reaction: the process has been in commercial use for more than a half century, catalysts have been highly optimized, yet the fundamental question of the nature of the essential oxygen species has not been conclusively resolved.

Recent fundamental studies, particularly those on well-characterized samples such as single crystals, have brought us much closer to an understanding of the mechanisms of the individual reactions and intermediates involved in this process. It is clear that the low-coverage atomic oxygen species most accessible in UHV studies are strongly nucleophilic in character, and react selectively with ethene. Epoxidation requires electrophilic oxygen species, as was recognized implicitly in the suggestion that molecular oxygen species, particularly adsorbed super oxides, are responsible for this reaction. The best evidence currently available is that these electrophilic species are not molecular, but are unstable atomic species formed on an electron-deficient silver surface, i.e., one containing substantial quantities of subsurface oxygen. Promotion with halides may serve to mimic the subsurface oxide, again leading to the formation of less-charged surface oxygen species. Given the renewed interest in the mechanism of epoxidation, perhaps direct evidence for the formation of electrophilic atomic oxygen will be obtained more readily than that for the often proposed molecular species, which have only recently been directly observed.

REFERENCES

1. **Thomas, C. L.,** *Catalytic Processes and Proven Catalysts,* Academic Press, New York, 1970, 214.
2. **Lefort, T. E.,** French Patent 729, 952, 1931.
3. **Wimer, W. E. and Feathers, R. E.,** *Hydrocarbon Process,* 215, November, 1975.
4. **Berty, J. M.,** in *Applied Industrial Catalysis,* Vol 1, Leach, B. E., Ed., Academic Press, New York, 1983, chap. 8.
5. **Voge, H. H. and Adams, C. R.,** *Adv. Catal.,* 17, 151, 1967.
6. **Kilty, P. A. and Sachtler, W. M. H.,** *Catal. Rev. Sci. Eng.,* 10, 1, 1974.
7. **Barteau, M. A. and Madix, R. J.,** in *The Chemical Physics of Solid Surfaces and Heterogeneous Catalysis,* King, D. A. and Woodruff, D. P., Eds., Elsevier, Amsterdam, 1982, chap. 4; and references cited therein.
8. **Clayton, R. W.,** *Chem. Soc. Spec. Period. Rep. Catal.,* 3, 70, 1980.
9. **Verykios, X. E., Stein, F. P., and Coughlin, R. W.,** *Catal. Rev. Sci. Eng.,* 22, 197, 1980.
10. **Sachtler, W. M. H., Backx, C., and Van Santen, R. A.,** *Catal. Rev. Sci. Eng.,* 23, 127, 1981.
11. **Carra, S. and Forzatti, P.,** *Catal. Rev. Sci. Eng.,* 15, 1, 1977.
12. **Spencer, N. D. and Somorjai, G. A.,** *Rep. Prog. Phys.,* 46, 1, 1983.
13. **Ostrovskii, V. E., Kulkova, N. V., Lopatin, V. L., and Temkin, M. I.,** *Kinet. Katal.,* 3, 160, 1962.
14. **Spath, H. T., Mayer, K., and Torkar, K.,** *J. Catal.,* 35, 100, 1975.
15. **Mross, M. D., Koopmann, J., Vogt, V., and Schwarzmann, M.,** German Patent 2,933,950, 1981.
16. **Metcalf, D. L. and Harriot, P.,** *Ind. Eng. Chem. Process Des. Dev.,* 11, 478, 1972.
17. **Rebsdat, S., Mayer, S., and Alfransedar, J.,** *Chemie Ing. Tech.,* 53, 850, 1981.
18. **Toyoda, Y., Ikeda, Y., and Kobura, Y.,** German Patent 2,940,480, 1980.
19. **Cannon, J. C.,** U.S. Patent 4,235,757, 1980.
20. **Kajimoto, T.,** European Patent 11,356,1980.
21. **Wernli, W. L., Fry, W., and Jonda, S. F.,** U.S. Patent 4,248,741, 1981.
22. **Cavitt, S. B.,** German Patent 2,951,970, 1980.
23. **Cavitt, S. B.,** U.S. Patent 4,206,128, 1980.
24. **Hayden, P., Spencer, C. B., Andrew, S. P. C., and Denny, P. J.,** British Patent 1,571,123, 1980.
25. **Nielson, R. P. and Jecminek, A. A.,** U.S. Patent 4,267,073, 1981.
26. **Khoobiar, S.,** U.S. Patent 4,169,099, 1979.
27. **Mitsuhata, M., Kumazawa, T., Kiguchi, I.,** *Jpn. Kokai Tokyo Koho,* 13(79), 485, 1979.
28. **Titzenthaler, E. and Schwen, R.,** German Patent 2,844,402, 1980.
29. **Worbs, H.,** Dissertation Technische Hochshule, Breslau; U.S. Office of Technical Service P. B. Rep. 98705, 1942.
30. **Twigg, G. H.,** *Trans. Faraday Soc.,* 42, 284, 1946.
31. **Hayes, K. E.,** *Can. J. Chem.,* 38, 225, 1960.
32. **Force, E. L. and Bell, E. T.,** *J. Catal.,* 40, 356, 1975.
33. **Gerei, S. V., Kholyavenko, K. M., and Rubanik, M. Y.,** *Ukr. Khim. Zhur.,* 31, 449, 1965.
34. **Cant, N. W. and Hall, W. K.,** *J. Catal.,* 52, 89, 1978.
35. **Kilty, P. A., Rol, N. C., and Sachtler, W. M. H.,** *Proc. 5th Int. Congr. Catal.,* Vol 2, Highwater, J., Ed., 1973, 929.

36. **Herzog, W.,** *Ber. Bunsenges. Phys. Chem.*, 74, 216, 1970.
37. **Charmon, H. B., Dell, R. M., and Teale, S. S.,** *Trans. Faraday Soc.*, 59, 453, 1963.
38. **Clarkson, R. and Cirillo, A.,** *J. Catal.*, 33, 392, 1974.
39. **Sato, N. and Seo, M.,** *J. Catal.*, 24, 224, 1972.
40. **Kobayashi, Y. M. and Kobayashi, H.,** *Proc. 6th Int. Congr. Catal.*, Chemical Society, London, 1976, 336.
41. **Egashira, M., Kuczkowski, R. L., and Cant, N. W.,** *J. Catal.*, 65, 297, 1980.
42. **Backx, C., De Groot, C. P. M., and Biloen, P.,** *Appl. Surf. Sci.*, 6, 256, 1980.
43. **Wachs, I. E. and Klemen, S. R.,** Proc. 7th Int. Congr. Catal., Tokyo, 1980, paper A48.
44. **Barteau, M. A. and Madix, R. J.,** *Surf. Sci.*, 103, L171, 1981.
45. **Huang, Y. Y.,** *J. Catal.*, 61, 461, 1980.
46. **Van Santen, R. A., Moolhuysen, J., and Sachtler, W. M. H.,** *J. Catal.*, 65, 478, 1980.
47. **Backx, C., Moolhuysen, J., Geenen, P., and Van Santen, R. A.,** *J. Catal.*, 72, 364, 1981.
48. **Van Santen, R. A. and De Groot, C. P. M.,** *J. Catal.*, 98, 530, 1986.
49. **Benton, A. F. and Drake, L. C.,** *J. Am. Chem. Soc.*, 56, 255, 1934.
50. **Sandler, Y. L. and Durigon, D. D.,** *J. Phys. Chem.*, 69, 4201, 1965.
51. **Czanderna, A. W.,** *J. Phys. Chem.*, 68, 4201, 1964.
52. **Buttner, F. H., Funk, E. R., and Udin, H.,** *J. Phys. Chem.*, 56, 657, 1952.
53. **Kummer, J. T.,** *J. Phys. Chem.*, 63, 410, 1959.
54. **Bagg, J. and Bruce, L.,** *J. Catal.*, 2, 93, 1963.
55. **Kollen, W. and Czanderna, A. W.,** *J. Colloid Interface Sci.*, 38, 152, 1972.
56. **Czanderna, A. W., Chen, S. C., and Biegen, J. R.,** *J. Catal.*, 33, 163, 1974.
57. **Czanderna, A. W.,** *J. Vac. Sci. Technol.*, 14, 408, 1976.
58. **Ackern, R. J. and Czanderna, A. W.,** *J. Catal.*, 46, 109, 1977.
59. **Czanderna, A. W.,** *J. Phys. Chem.*, 70, 2120, 1966.
60. **Biegen, J. R. and Czanderna, A. W.,** *J. Vac. Sci. Technol.*, 8, 376, 1971.
61. **Czanderna, A. W., Frank, O., and Schmidt, W. A.,** *Surf. Sci.*, 38, 129, 1973.
62. **Smeltzer, W. W., Tollefson, E. L., and Cambron, A.,** *Can. J. Chem.*, 34, 1046, 1956.
63. **Rovida, G., Pratesi, F., and Ferroni, E.,** *J. Catal.*, 41, 140, 1976.
64. **Czanderna, A. W.,** *Thermochim. Acta*, 24, 359, 1978.
65. **Obstrovskii, V. and Temkin, M.,** *Kinet. Catal.*, 7, 466, 1966.
66. **Richardson, P. C. and Rossington, D. R.,** *J. Catal.*, 20, 420, 1971.
67. **Sachtler, W.,** *Catal. Rev.*, 4, 27, 1970.
68. **Rovida, G. and Pratesi, F.,** *Surf. Sci.*, 52, 542, 1975.
69. **Rovida, G., Ferroni, E., Maglietta, M., and Pratesi, F.,** in *Adsorption Desorption Phenomena*, Ricca, F., Ed., Academic Press, London, 1972, 417.
70. **Wachs, I. E. and Klemen, S. R.,** *J. Catal.*, 68, 213, 1981.
71. **Grant, R. B. and Lambert, R.,** *Surf. Sci.*, 146, 256, 1984.
72. **Campbell, C. T.,** *J. Catal.*, 94, 436, 1985.
73. **Engelhardt, H. A. and Menzel, D.,** *Surf. Sci.*, 57, 591, 1976.
74. **Rovida, G., Pratesi, F., Maglietta, M., and Ferroni, E.,** *Surf. Sci.*, 43, 230, 1974.
75. **Bowker, M., Barteau, M. A., and Madix, R. J.,** *Surf. Sci.*, 92, 528, 1980.
76. **Barteau, M. A. and Madix, R. J.,** *Surf. Sci.*, 97, 101, 1980.
77. **Rovida, G.,** *J. Phys. Chem.*, 80, 150, 1976.
78. **Albers, H., Droog, J. M. M., and Bootsma, G. A.,** Surf. Sci., 64, 1, 1977.
79. **Albers, H., Van der Wal, W., and Bootsma, G. A.,** *Surf. Sci.*, 68, 47, 1977.
80. **Albers, H., Van der Wal, W., Gijzeman, O. L. J., and Bootsma, G. A.,** *Surf. Sci.*, 77, 1, 1978.
81. **Bradshaw, A. M., Menzel, D., and Steinkilberg, S.,** *Faraday Discuss.*, 58, 46, 1974.
82. **Barteau, M.,** Ph.D. dissertation, Stanford University, Palo Alto, Calif., 1981.
83. **Fuggle, J. C. and Menzel, D.,** *Surf. Sci.*, 53, 21, 1975.
84. **Rao, C. N. R., Kamath, P. V., and Yashonath, S.,** *Chem. Phys. Lett.*, 88, 13, 1982.
85. **Heiland, W., Iberl, F., and Tanglauer, H.,** *Surf. Sci.*, 53, 383, 1975.
86. **Backx, C., De Groot, C. P. M., and Biloen, P.,** *Surf. Sci.*, 104, 300, 1981.
87. **Sexton, B. A. and Madix, R. J.,** *Chem. Phys. Lett.*, 76, 294, 1980.
88. **Campbell, C. T. and Paffett, M. T.,** *Surf. Sci.*, 143, 517, 1984.
89. **Campbell, C. T.,** *J. Vac. Sci. Technol.*, 2, 1024, 1984.
90. **Bradshaw, A. M., Engelhardt, H. A., and Menzel, D.,** *Ber. Bunsenges Phys. Chem.*, 76, 500, 1972.
91. **Engelhardt, H. A., Bradshaw, A. M., and Menzel, D.,** *Surf. Sci.*, 40, 410, 1973.
92. **Barteau, M. A. and Madix, R. J.,** *J. Chem. Phys.*, 74, 4144, 1981.
93. **Bowker, M.,** *Vacuum*, 33, 669, 1983.
94. **Backx, C., De Groot, C. P. M., Biloen, P., and Sachtler, W. M. H.,** *Surf. Sci.*, 128, 81, 1983.

95. **Eichmans, J., Goldman, A., and Otto, A.,** *Surf. Sci.,* 127, 153, 1983.
96. **Prince, K. C. and Bradshaw, A. M.,** *Surf. Sci.,* 126, 49, 1983.
97. **Joyner, R. W. and Roberts, M. W.,** *Chem. Phys. Lett.,* 60, 459, 1979.
98. **Grant, R. B. and Lambert, R.,** *J. Catal.,* 92, 364, 1985.
99. **Campbell, C. T. and Paffett, M. T.,** *Surf. Sci.,* 139, 396, 1984.
100. **Kitson, M. and Lambert, R.,** *Surf. Sci.,* 100, 368, 1980.
101. **Goddard, P. J. and Lambert, R. M.,** *Surf. Sci.,* 67, 180, 1977.
102. **Tu, Y. Y. and Blakely, J. M.,** *J. Vac. Sci. Technol.,* 15, 563, 1978.
103. **Bowker, M. and Waugh, K. C.,** *Surf. Sci.,* 134, 639, 1983.
104. **Bowker, M. and Waugh, K. C.,** *Surf. Sci.,* 155, 1, 1985.
105. **Campbell, C. T. and Paffett, M. T.,** *Appl. Surf. Sci.,* 19, 28, 1984.
106. **Campbell, C. T. and Koel, B. E.,** *J. Catal.,* 92, 272, 1985.
107. **Campbell, C. T.,** *J. Catal.,* 99, 28, 1986.
108. **Marbrow, M. and Lambert, R.,** *Surf. Sci.,* 65, 314, 1977.
109. **Briggs, D., Marbrow, R., and Lambert, R.,** *Surf. Sci.,* 65, 314, 1977.
110. **Goddard, P. and Lambert, R.,** *Surf. Sci.,* 107, 519, 1981.
111. **Kitson, M. and Lambert, R.,** *Surf. Sci.,* 109, 60, 1981.
112. **Kitson, M. and Lambert, R.,** *Surf. Sci.,* 110, 205, 1981.
113. **Grant, R. B. and Lambert, R.,** *J. Catal.,* 93, 92, 1985.
114. **Campbell, C. T.,** *J. Phys. Chem.,* 89, 5789, 1985.
115. **Spencer, N. D. and Lambert, R.,** *Surf. Sci.,* 104, 63, 1981.
116. **Neilsen, R. and La Rochelle, J.,** U.S. Patent 4,356,312, 1982.
117. **M. Murray,** *Aust. J. Sci. Res.,* A3, 433, 1950.
118. **Wan, S.-W.,** *Ind. Eng. Chem.,* 45, 234, 1953.
119. **Forgani, F., and Montarnal, R.,** *Rev. Inst. Fr. Pet.,* 14, 191, 1959.
120. **Alfani, F. and Carberry, J. J.,** *Chim. Ind.,* 52, 1192, 1970.
121. **Kenson, R. E. and Lapkin, M.,** *J. Phys. Chem.,* 74, 1493, 1970.
122. **Klugherz, P. D. and Harriot, P.,** *AIChE J.,* 17, 856, 1971.
123. **Verma, A. and Kaliaguine, S.,** *J. Catal.,* 30, 430, 1973.
124. **Kripyylo, P., Mogling, L., Ehrchen, H., Harkanyi, I., Klose, D., and Beck, L.,** *Chem Tech.,* 31, 82, 1979.
125. **Dettwiler, R. H., Baiker, A., and Richarz, V.,** *Helv. Chim. Acta.,* 62, 1680, 1979.
126. **Dweydari, A. W. and Mee, C. H. B.,** *Phys. Status Solidi,* 17, 247, 1973.
127. **Bowker, M.,** *Surf. Sci.,* 100, L472, 1980.
128. **Barteau, M. and Madix, R. J.,** *Surf. Sci.,* 140, 108, 1984.

Chapter 5

CATALYTIC HYDROGENATION OF CARBON MONOXIDE: METHANE AND METHANOL SYNTHESIS

I. GENERAL

The production of both methane and methanol from the hydrogenation of CO has received new attention in the past few years. This is due both to potential uncertainty on the oil supply market and to the added increase in the economic attractiveness of producing synthetic natural gas (SNG) from hydrogen-deficient materials. The liquefaction of coal and subsequent gasification of char holds the most promise, partly because technology has existed for more than half a century, at least in a rudimentary stage, and partly because it is the most available, with large deposits of coal easily accessible. Conversion of fossil fuels to gaseous products, a mixture of methane, carbon monoxide, and hydrogen, can be accomplished by various processes. The synthesis gas can then be catalytically converted to various organic products including methane, alcohols, and higher hydrocarbons.

II. HISTORY

The commercial exploitation of synthesis gas has a distinguished history. The formation of methane from a mixture of carbon monoxide and hydrogen over a nickel catalyst was first discovered by Sabatier and Senderens[1] in 1902. Methanol synthesis from CO and H_2 was first reported by Patart[2] in 1921, and in 1923 Badische Aniline and Soda Fabrik (BASF) patented a process for the selective hydrogenation of CO to methanol using a catalyst composed of oxides of zinc and chromium.[3] Synthesis plants employing these catalysts operated at 250 to 350 atm and temperatures between 600 and 680 K, and thus were considered high-pressure methanol synthesis plants.

During 1923 to 1926, Fischer and Tropsch[4,5] developed the normal pressure process for the conversion of synthesis gas to liquid hydrocarbons using a catalyst consisting of iron and cobalt with copper as promoter. The process then underwent several modifications which included the addition of ThO_2 to the catalyst as well as an increase in pressure to 10 to 15 atm. The nonselective generation of organic compounds through the hydrogenation of CO has become known as Fischer-Tropsch (F-T) synthesis.

Following fundamental research by Fischer and Tropsch, interest in CO hydrogenation grew in many countries almost simultaneously, but with the availability of cheap oil, research was virtually halted in all countries, except South Africa, by the late 1950s. The only commercial plant currently in operation is SASOL plant in South Africa; in fact, by 1982 they had three plants in operation.[6] The commercial F-T synthesis employs the iron-based catalysts. The main products are found to be olefins, paraffins, water, and carbon dioxide with minor amounts of alcohols and other oxygenated compounds.

The continuing uncertainty in oil prices and instability in the oil supply market were the incentives for synthetic fuel research to begin again in the 1970s and is presently being actively continued on both fundamental and technological levels in many countries. Specific topics have ranged from catalyst preparation, mechanism studies, kinetic studies, and pilot-plant operation using different reactor configurations. The reaction mechanisms of the F-T synthesis have been reviewed by Mills and Steffgen,[7] Vannice,[8] Ponec,[9] Bell,[10] and Anderson.[11] Bell[10] proposed a mechanism of F-T synthesis type, explaining the synthesis of both the hydrocarbons and the higher alcohols. His idealized mechanism leads to the formation of a Schultz-Flory distribution of molecular weights, which is often encountered in polymerization reactions.

Although the catalytic hydrogenation of the oxides of carbon is actually part of the more generalized reaction systems associated with the Fischer-Tropsch synthesis, different process operating conditions used both in methane and methanol formation, and appropriate choice of the catalyst, as compared to those used in F-T synthesis suggest that they be treated as a separate and distinct subject. Selective catalysts do exist for methane and methanol syntheses, and surprisingly, these were the original CO hydrogenation catalysts to be formulated. These cover selective processes of the greatest industrial interest, and naturally have received the most attention. In fact, when largely sulfur-free synthesis gases became available to be used about 1970 in conjunction with highly active copper catalysts, the stringent reaction conditions for methanol synthesis could be relaxed.

This chapter is devoted to both methanation and methanol syntheses, beginning with an introductory statement of the reactions, catalysts, processes, and reaction kinetics and following with detailed accounts of catalytic chemistry. A coherent picture of catalyst structure and surface reaction mechanism has begun to emerge only in the last few years. To a large extent, and especially for the methanation, this understanding has arisen through the use of newly developed surface techniques typically utilizing well-defined, low surface-area materials in UHV conditions, as emphasized in the previous discussions.

III. METHANATION

A. Reactions, Catalyst, and the Process

Methanation, the hydrogenation of CO to CH_4, converts a low BTU synthesis gas mixture into a clean and high BTU substitute natural gas (SNG). Industrially, methanation is carried out almost exclusively with Raney® Ni or supported Ni systems (kieselguhr, Al_2O_3, and SiO_2) but iron, cobalt, ruthenium, rare earth intermetallics,[21] and Ni-Fe[12] also work. Vannice[8] has reported selectivities for the methanation reaction over the group VIII metals to be in the following decreasing order: Pd > Pt > Ir > Ni > Rh > Co > Fe > Ru.

The methanation reaction can be described by

$$CO + 3H_2 = CH_4 + H_2O \qquad (1)$$

$$\Delta H = -221 \text{ kJ/mol}$$

When CO and H_2O are present at conditions for Reaction 1, the water-gas shift reaction may also occur:

$$CO + H_2O \rightleftharpoons H_2 + CO_2 \qquad (2)$$

$$\Delta H = -40 \text{ kJ/mol}$$

Using nickel catalyst, some ethane, usually <5%, forms under normal methanation conditions.[13] The other reaction which is also of practical significance is:

$$2CO = CO_2 + C \qquad (3)$$

$$\Delta H = -173 \text{ kJ/mol}$$

This disproportionation of CO, also called Boudouard reaction, can lead to deposition of carbon on the catalyst with eventual resultant deactivation. Therefore, water is purposely added to reduce carbon formation in the reactor and to generate hydrogen by the water-gas shift reaction.

Methanation is typically carried out between 500 to 700 K and pressures exceeding 20 atm but up to 100 atm and it is essentially irreversible under these conditions. The turnover rates vary between 10^{-2} to 10 methane molecules per surface site per second.[14] Temperatures higher than 700 K can result in severe catalyst deactivation due to metal sintering.[8]

The free-energy values of all these reactions have large negative values over a wide range of temperatures. There are no thermodynamic limitations on the methane yield at atmospheric pressure and temperatures between 500 and 700 K when the molar ratio H_2/CO in the synthesis gas is larger than 5.0. The effect of the operating conditions on equilibrium yield of methane has been discussed by Mills and Steffgen.[7]

Since the methanation reaction is highly exothermic, the heat removal system must be able to withstand the heat release, keeping the temperature inside the reactor system constant within narrow limits. As a result, several types of catalytic methanation reactors have been studied in order to achieve maximum reactor performance in terms of conversion, selectivity, and life of catalyst:

1. Fixed bed, cooled by heat exchange surfaces (Dirksen and Linden[15])
2. Recycle reactor (Seglin et al.[16] and Fitzharris and Katzer[22])
3. Fluidized bed, cooled indirectly by heat-exchange surfaces (Cobb and Streeter[17])
4. Slurry reactor (Blum et al.[18])
5. Tube wall, with catalyst supported on and cooled by heat-exchanger tubes (Haynes et al.[19] and Goyal[20])

The tube-wall reactors have received special attention because of their simple construction and excellent temperature control. The reactor walls are coated with the catalyst material by means of the flame- or plasma-spraying techniques.[20] Haynes et al.[19] have shown that the performance of Raney® nickel-coated wall reactors is significantly better than that of precipitated nickel catalyst pellets; the catalyst-sprayed plates resulted in higher production of methane based on a similar amount of catalyst, longer catalyst life, or lower rate of deactivation, lower concentration of carbon monoxide in the product gas, and lower pressure drop across the reactor.

The product gases from the different types of gasifiers require essentially the same type but different extents of treatment. The purification steps are relatively well established, but there is some uncertainty about the behavior of processes for removal of sulfur compounds. Sulfur poisons nickel catalyst; it is usually assumed that the H_2S content of the feed gas must be below about 0.1 ppm. Sulfur apparently bonds so strongly to metal surfaces that marked activity reduction takes place even at extremely low gas-phase concentrations of sulfur-containing compounds. Regeneration of the catalyst is usually impossible or impractical due to the essentially irreversible adsorption of sulfur compounds on metals, and has been a constant concern of the industry. An understanding of the mechanism and nature of sulfur adsorption would certainly provide a basis for relating the changes in the activity of methanation catalysts. This subject is considered further in a later section.

In the nickel-catalyzed methanation reaction, catalyst deactivation by carbon deposition often is a severe problem as it is in F-T synthesis. The surface carbon also plays a major role in the synthesis. Kelley and Goodman[27] have classified the phases involved in the methanation reaction according to their function; the active carbon species is termed as 'carbidic' and the inactive is designated as 'graphitic'. Thus, if the process is to have long-term operation at steady state without frequent catalyst replacement, it is essential to find additives which not only inhibit or retard this transformation, but also develop a tolerance to low concentrations of sulfur.

Several patents devoted to improve yields of methane have been issued. Addition of rhodium, potassium, ruthenium, and/or iridium to nickel have been shown to improve the selectivity and it is claimed that Ir has good resistance against poisoning by sulfur (Figure

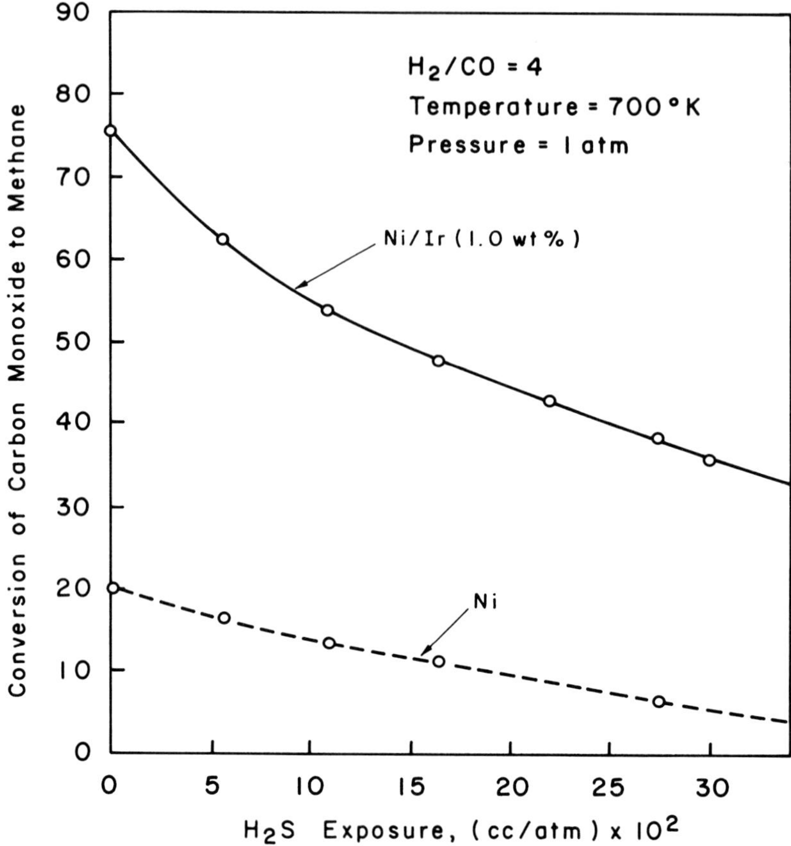

FIGURE 1. Effect of the iridium promoter on the methanation activity and hydrogen sulfide poisoning of nickel on alumina (25% Ni) catalysts.[23]

1).[23] The heats of adsorption of both hydrogen and CO increase by the addition of potassium to nickel surfaces which may alter their relative surface concentration as well as their dissociation probability.[14] However, the function and disposition of alkali metals are far from clear, and further studies are needed to explain their mechanistic role in methanation reaction.[24]

It is evident that this synthesis also forms a complex system with possibly six, or more, reactive gas-phase components, and intricate metal-alkali-carbon interactions all interfering in the synthesis. By separating out some of the components, the following sections will attempt to present the results which have been obtained under less well-controlled conditions, followed by the studies conducted on single crystals.

B. Mechanism of CO-H_2 Interaction: An Overview

The renewed interest in catalytic methanation has brought with it a number of investigations intended to explain the reaction mechanism of CO hydrogenation. It is also pertinent to consider here the important results obtained under conditions for the synthesis of higher hydrocarbons and alcohols because of their importance to the subsequent discussion of the methanation mechanism, especially in the initial stages. The major question proposed in the literature until recently was whether or not nickel carbonyl or nickel carbide is formed as an intermediate in CO reduction.

Of the reviews and books published so far on methanation[7-11,14,25-28] those by the following are particularly noteworthy: Mills and Steffgen(1973),[7] Ponec(1978),[9] Somorjai(1981),[14] and Kelley and Goodman(1982).[27]

The first technique used to elucidate the mechanism of CO hydrogenation was volumetric adsorption. These studies[29-31] with iron catalysts have suggested that a surface complex of the stoichiometry H_2CO may be an important intermediate in CO reduction. Kolbel and Roberg[29] reported an enhancement in CO/H_2 adsorption for iron, and, at room temperature, this ratio was near unity. Kolbel and Hanus[32] also analyzed the species which desorbed from the iron species. The results showed the presence of H_2CO species. Subsequent IR studies[33] indicated the presence of carbonyl, carbonate, and carboxyl species on the iron surface. Blyholder and Neff,[34] in coadsorption experiments on Fe-SiO_2 catalysts using IR techniques, also favored the formation of a surface complex H_xCO. However, the same workers failed to observe any chemisorbed alcohol species in their IR studies of CO/H_2 adsorption on nickel. Similarly, Madey et al.[35] failed to find evidence for oxygenated species on nickel.

Mutual enhancement of H_2 and CO was also reported for metal films. However, no reaction products could be found in the gas phase as a result of the simultaneous adsorption of CO and H_2 on nickel films in the temperature range 77 to 353 K.[36] The capacity of the films for CO uptake was considerably increased at 353 K by the presence of H_2. Horgan and King,[37] in coadsorption experiments using UHV techniques, reported such an effect in the range of 300 to 650 K. At 273 K, however, this effect was not seen.[36] Additionally, when the total coverage exceeds the monolayer coverage based on H_2, H_2 is displaced into the gas phase by CO.[36,38]

In contrast to the formation of a surface complex with the stoichiometry H_2CO, other researchers[39,40] proposed that reaction proceeds through a surface carbide intermediate, the carbide being formed from the decomposition of adsorbed CO before hydrogenation occurs. Araki and Ponec[39] performed the experiments with evaporated Ni films in a static UHV apparatus. When a reaction mixture was admitted to the fresh film, CO_2 appeared first, before CH_4 was formed (Figure 2), indicating that CO_2 was formed by disproportionation (3).

Joyner and Roberts[41] have also shown in their XPS and UPS studies that CO can be dissociated on a number of metal surfaces (such as tungsten, molybdenum, nickel, cobalt, iron, etc.) at temperatures below 420 K. Surface carbon thus obtained is then hydrogenated to methane. In fact, subsequently, Joyner[42] has observed that the interaction of methanol with nickel at 300 K resulted in the dissociation into an adsorbed layer of CO.

Using quantum calculations (extended Debye-Huckel theory), Kolbel and Tillmetz[43] computed the stabilities of surface species of the type $\mathrm{M{-}\overset{\displaystyle H}{\overset{|}{R}}{-}OH}$, where M is a metal atom (Fe, Co, or Ni) and R is an alkyl group. Their results indicated that species containing alkyl groups bonded to oxygen are unstable on nickel and should decompose into methene structures on the nickel surface:

$$\mathrm{Ni{-}\underset{\displaystyle H}{\underset{|}{R}}{-}OH \rightarrow Ni{-}R + H_2O} \quad (4)$$

Therefore, hydrogenation of CO on a nickel surface most probably propagates through a carbide or alkyl surface.

Although the carbide as an intermediate in methanation was unacceptable to Vlasenko and Yuzefovich[44] and Farrauto[45] on the basis that the rate of methanation exceeds that of carbide formation, this view misrepresents the observations in two ways: (1) formation of surface carbon does not necessarily follow the same kinetics as those followed by metal

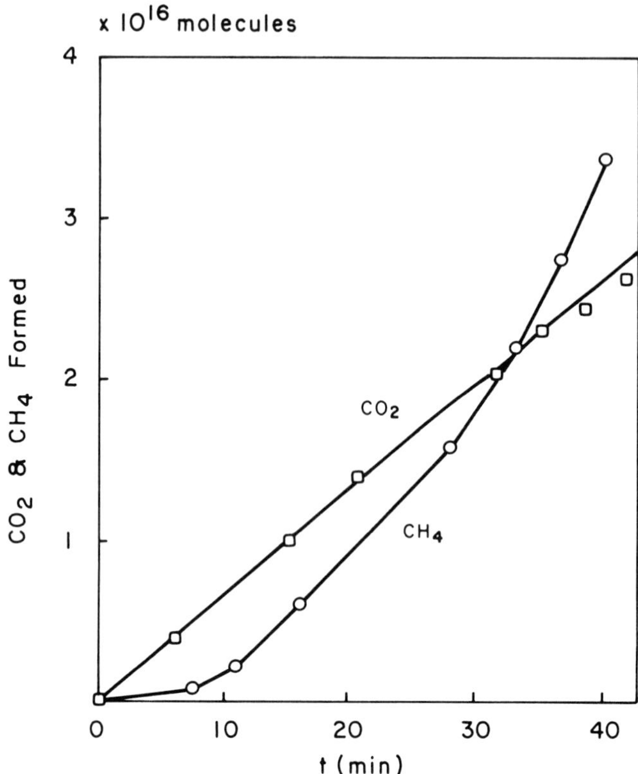

FIGURE 2. Formation of CO_2 and CH_4 from the reaction mixture H_2/CO as a function of time over Ni film at 523 K. (From Araki, M. and Ponec, V., *J. Catal.*, 44, 439, 1976. With permission.)

carbiding and (2) H_2 in methanation process removes surface carbon as methane thus making more surface area available for CO chemisorption.[25] Further supportive evidence for the carbide mechanism was given by Ertl and co-workers[46,47] as well as by Rewich and Wise.[48]

By way of summary of the foregoing discussion (prior to the late 1970s) of this reaction, the following conclusions drawn from the two extended reviews[7,9] are described because of their importance to the subsequent discussion of low surface-area-catalyzed methanation reaction.

In 1973, Mills and Steffgen[7] summarized all suggestions in the literature into essentially three different mechanisms:

1. CO dissociates and forms 'carbides'. These are particularly hydrogenated into CH_x which may be either hydrogenated to produce methane or may polymerize into a hydrocarbon chain.
2. The essential intermediate contains oxygen as HCOH.
3. Chain propagation in the second mechanism proceeds by carbonyl insertion.

On the other hand, Ponec,[9] in 1978, pointed out that there are problems with each of the three mechanisms proposed in the literature when they are considered separately. For example, carbonyls (Mechanism [3]) are unstable at the temperature usually used. However, it is known that when F-T synthesis is carried out on surfaces carbided with radioactive-labeled carbon, methane contains most of the radioactivity with little in higher hydrocarbons which indicates that the formation of methane and higher hydrocarbons may involve different intermediates.[49] Ponec suggested that a combination of two of them can explain the data, and accordingly, proposed the following mechanism:

1. CO dissociates and surface carbon is partially hydrogenated to produce CH_x species.
2. CH_x is hydrogenated to form methane.
3. CH_x grows into a longer chain by CO insertion and partial hydrogenation.

It is to be noted that the above reviews dealt primarily with results obtained on high surface-area catalysts. Two additional general reviews on F-T synthesis have appeared recently.[169,170]

It is obvious that evidence for enolic species is not very strong and also has been recently questioned by King.[50] In an *in situ* IR study over Ru/SiO_2, Ru/Al_2O_3, and Fe/SiO_2, he saw no evidence of any active species other than the surface species CO, carbon, and hydrogen. The only spectral species he observed were those for the structures containing CH_2 and CH_3 groups. These species were concluded to be inactive in the formation of light hydrocarbon products.[50,51] On the other hand, Tamaru[52] has suggested that the observed hydrocarbon species can undergo hydrogenolysis, and therefore may act as a reservoir for CH_x groups. Biloen and Sachtler[169] have suggested that the active carbide is a heterogeneous mixture of C(a), CH(a), and other hydrocarbon intermediates which can undergo both hydrogenation and polymerization to form various hydrocarbon products.

Temperature-programmed surface reaction (TPSR) and temperature-programmed desorption (TPD) have also been used successfully to study CO methanation on silica-supported nickel[52] and ruthenium.[53] In these studies, the reactive carrier gas was kept at constant partial pressure, and the catalysts initially contained a monolayer of CO adsorbed under nonreactive conditions at room temperature. For Ni/SiO_2, single CH_4 and H_2O TPSR peaks, each with a peak temperature of 473 K, were observed (Figure 3). The simultaneous appearance of both products suggests that the slow step in the mechanism is the activation of adsorbed CO. The desorption and reaction spectra for Ru/SiO_2 were similar to those for Ni/SiO_2.

Evidence thus far has indicated that CO hydrogenation over nickel proceeds through a carbon surface species, rather than an oxygenated surface intermediate. Therefore, the structure of surface carbon intermediates is fundamental to the activity and selectivity of CO hydrogenation catalysis by transition metals.

C. Reactivity of Carbon

The role of surface carbon has recently been studied on supported Ni catalysts as well as on Ni films.[39-41,55-64] The active carbon overlayer that forms on nickel during the synthesis reaction maintains its activity to produce methane only in a rather narrow temperature range. Wentreck et al.[40] have shown that at temperatures above 720 K, the reactive 'carbidic' carbon deposits begin to graphitize into an unreactive surface carbon deposit, even in the presence of excess hydrogen, thus making 700 K as an upper limit for CO hydrogenation.[55,56] Below 450 K, the dissociation rate of CO to produce the active carbon that is to be rehydrogenated to methane is too slow on most transition-metal surfaces. Geus and co-workers[64] further narrowed down this temperature range: the rate of reaction of hydrogen with the carbon, deposited from CO on Ni/SiO_2, passed through a maximum at 480 K and dropped steeply above 580 K. This dissociated carbon also led to the formation of a bulk nickel carbide as well as dissolution of carbon interstitially.

Bartholomew,[60] in 1982, directed his review at deactivation phenomena connected with methanation. He concluded that the characteristic reactivity of various carbon states was independent of the source of carbon but a strong function of the temperature and composition of the exposing gas. Subsequently, Bartholomew and Vance[63] also showed that the activation energy for carbon hydrogenation on nickel was independent of support (3% Ni on SiO_2, Al_2O_3, or TiO_2).[63] While these authors also reported that specific rates of carbon hydrogen-

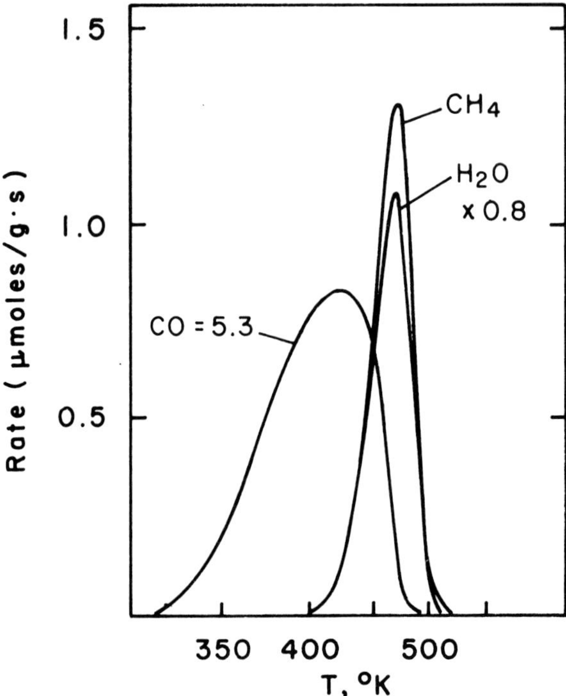

FIGURE 3. TPSR of CO adsorbed on Ni/SiO_2 reacting with H_2. (From Falconer, J. L. and Zagli, E., *J. Catal.*, 62, 280, 1980. With permission.)

ation varied as much as a factor of 10, Ozdogan et al.,[75] in their temperature-programmed reaction studies, determined that these rates were independent of support. The later measurements were, however, made at high carbon coverages on catalysts of moderately high nickel loading (16%). Several carbon states have also been identified on supported iron catalysts during Fischer-Tropsch synthesis which exhibit characteristic reactivities toward hydrogen.[61,62]

It has also been reported that active carbon that forms from CO disproportionation on nickel surfaces reacts readily with water to form CO_2 and CH_4:[14,76]

$$2C + 2H_2O = CO_2 + CH_4 \tag{5}$$

At low carbon surface coverages, besides CO_2 and CH_4, H_2 is also produced.[64] This should be expected since a significant partial pressure of water is maintained in the industrial process to reduce carbon formation and to generate hydrogen.

It may also be noted that following adsorption on the metal surface of nickel and ruthenium, CO and CO_2 methanation proceed by the same mechanism.[53,54] These authors have shown that CO_2 first dissociates on the metal surface, which then reacts with hydrogen to form methane.

From the foregoing discussion, it is clear that the nature of CO dissociation on the surface, and in particular the character of the bonding of the surface carbon that deposits due to this dissociation, plays a prominent role in understanding of the methanation mechanism. Recent ultrahigh vacuum surface science investigations[14,27] on well-defined surfaces have corroborated many of the results attained under less well-controlled conditions and clarified several of the uncertainties which have been associated with this catalytic reaction system. However,

Table 1
KINETIC PARAMETERS FOR METHANATION OVER SUPPORTED Ni

N (sec^{-1})	CO^n	H_2^n	E(kJ/mol)	Ref.
0.08	−0.5— −0.2	0.6—0.8	113	62
0.09	(−)	(+)	117—130	63
0.06	−0.3	0.6	73—80	64
	−0.5— 0	0.15	104	65
0.04	—	—	84	66

before attempting to generalize a mechanism, we will review the forms of kinetic expressions proposed for this process. It must be pointed out that the kinetics and reactor studies are almost entirely qualitative since so much of the essential information remains proprietary.

D. Kinetic Models

Although the kinetics of methanation have been studied over transition metals, the most widely reported is that of nickel on alumina. Experimental conditions have varied in different studies over a wide range for the operating parameters, including pressure (1 to 70 atm), temperature (450 to 750 K), H_2/CO ratio (3 to 100), reactor type (fixed vs. fluidized bed), and reactor operation (integral vs. differential). Likewise, the kinetic models developed widely differ in their basic nature as well as their implications. In many cases, the range of experimental variables is too small to lead to a statistically meaningful rate expression.

Analyses of kinetic models were based on the 'carbide' theory and the 'enol' complex theory. In simple form, the carbide theory postulates that CO chemisorbs, and then dissociates on the catalyst surface. Surface-adsorbed carbon is then hydrogenated to form methane; adsorbed oxygen is removed as CO_2 and/or H_2O. On the other hand, Vlasenko and Yuzefovich[44] have suggested that the critical step is the formation of the formaldehyde enol complex. This complex is formed through the surface reaction of CO(a) and H_2(a). The general forms of the rate expressions are, however, qualitatively the same in both the theories for similar rate-controlling steps. Some general conclusions are apparent:

1. The methanation rate has a positive dependence on H_2 partial pressure.
2. The methanation rate decreases with increasing partial pressure of CO.
3. In most, though not all cases, reaction products do not inhibit methanation over nickel.

The values of the parameters in the form

$$r = k\, p_{H_2}^n\, p_{CO}^n \exp(-E/RT) \tag{6}$$

for selected references are presented in Table 1. The methanation turnover numbers, N, at 550 K are also included in this table.

Vannice[66] used different supported and unsupported nickel catalysts, and although apparent activation energy varied between 105 to 138 kJ/mol, most values were around 113 kJ/mol. The general consistency of the pressure dependencies is a strong indication that no major change in the microscopic reaction path occurs in the methanation reaction although nickel was dispersed on a number of support materials. Similarly, the correlation of the pressure dependence was obtained in the range 1 to 15 atm.[69] Even with the variation in activation energy, the turnover numbers agreed within a factor of about 2. The activation energy for methanation over Co-Al_2O_3 is reported to be 113 to 117 kJ/mol.[71,72] The methanation on supported ruthenium catalyst also has generally been reported to follow a negative order for CO and a positive order for hydrogen.[10,73,74]

It should be pointed out that in these kinetic analyses many proposals for the rate-limiting step have been offered. Some of these proposals are supported by kinetic data, suggesting the possibility that different surface steps may be controlling under certain conditions. Therefore, the discussion of the results becomes especially difficult when technical catalysts are used and too many parameters may influence the results. Nevertheless, the modern applications of kinetic measurements at process-like conditions with surface analytical techniques on well-defined systems (single crystals) have begun to combine the areas of modern UHV science and the surface science of heterogeneous catalysis, while at the same time recognizing the extreme complexity of the real industrial practice. In the following section, recent UHV results obtained on well-defined surfaces as model nickel catalysts will be considered. A major aim of most of these studies is the derivation of a mechanism describing the complex reaction in terms of individual steps.

E. Single-Crystal Studies

The hydrogenation of CO over well-defined surfaces of Ni, Ru, Rh, and W have been studied by a variety of modern surface techniques including TPD[77-81] and HREELS.[14,82] Much of the work has been conducted on W(100). TPD measurements on Ni(100) have shown the formation of a new surface species upon interacting CO with preadsorbed H(a).[77] However, due to kinetic limitations, no product detection was made at 10^{-3} atm pressure. The recent development of transfer techniques, high-pressure-UHV system, as described in Chapter 2, has, to some extent, alleviated this pressure-gap problem. Thus, the method now permits a direct comparison with kinetic measurements traditionally employed in the field of catalysis.

For the methanation reaction, several groups have combined surface analysis techniques with high pressure (up to six atm) kinetic measurements to study CO-H_2 interactions over transition metals.[27,28,65,83-87] By far, the most comprehensive work is that of Goodman, who has examined CO adsorption and desorption, carbide reaction in H_2, and CO-H_2 interactions on Ni surfaces,[27,28,77,86] as well as on Ru.[78,87,88] Several common observations have been found in these studies which illuminate the metal-CO interaction.

Unlike results for previous epoxidation of ethene studies, this heterogeneous reaction forms a strong linkage between classical catalytic research and ultrahigh vacuum catalysis. Figure 4 illustrates the Arrhenius plot of the methane synthesis rate over Ni(100) and Ni(111) at elevated partial pressures of CO and H_2[24,86] along with the dispersed Ni/Al_2O_3[66] and thin film Ni/Al_2O_3.[91] Excellent agreement between the kinetic results on the two crystal surfaces and the thin film model, and also with the highly dispersed supported nickel catalysts have been obtained, suggesting that the methanation reaction is structure insensitive.[28]

On the Ni(100) face, Goodman et al.[28,86] observed a low level of carbon species and the absence of oxygen. AES analysis indicated an active surface carbonaceous species to be a 'carbidic' form which can be produced in the absence of hydrogen by heating the crystal in pure CO at 600 K. Heating at higher temperatures results in a 'graphitic' carbon species. Figure 5 shows their AES spectrum along with single-crystal graphite and with bulk nickel carbide.[86] At high pressure, the 'carbidic' carbon forms have been shown to react with hydrogen to form methane. Quantitative aspects of the two carbon regions are furnished in Figure 6; the apparent saturation carbide level increases with temperatures between 500 and 700 K.[28] The saturation level of carbide at 600 K gave rise to p (2×2) LEED patterns, suggesting a coverage of 0.25 monolayer. It was also observed that the rate of carbide removal is a strong function of temperature and carbide coverage.

F. Mechanism of Methanation Over Nickel

The preceding background information, complemented by recent fundamental studies, has brought us much closer to an understanding of the mechanisms of the individual reactions

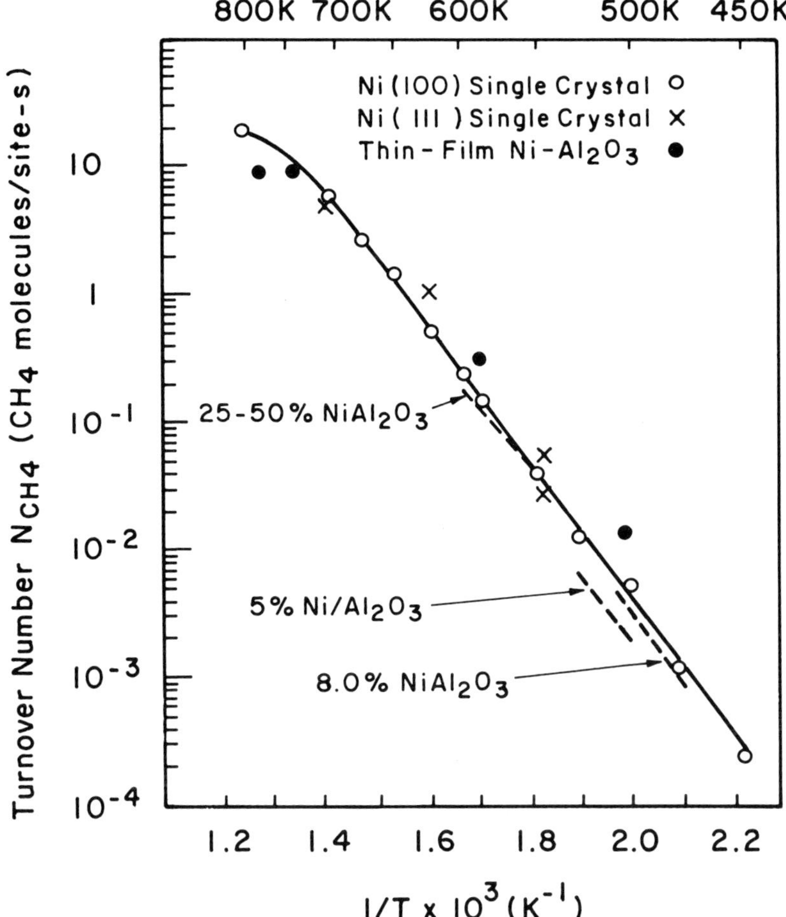

FIGURE 4. A comparison of the rate of methane synthesis over single-crystal nickel catalysts, thin film Ni-Al_2O_3 and supported Ni/Al_2O_3 catalysts. (From Bischke, S. D., Goodman, D. W., and Falconer, J. L., *Surf. Sci.*, 150, 351, 1985. With permission.)

and intermediates involved in this process. However, rate expressions for the elementary surface processes are virtually unknown at present.

We may assume here that all the faces of nickel contribute to the formation of methane. Since CH_4 is the main hydrocarbon product from Ni surfaces, it is clear that the paramount mechanism is the dissociation of CO, followed by the rehydrogenation of the surface carbon atoms to CH_4:

$$CO(a) \rightarrow C(a) + O(a) \tag{7}$$

$$C(a) + 4H(a) \rightarrow CH_4 \tag{8}$$

Reaction 8 represents a series of surface hydrogenation steps 9 to 12, as described below, which must occur:

$$C(a) + H(a) \rightarrow CH(a) \tag{9}$$

$$CH(a) + H(a) \rightarrow CH_2(a) \tag{10}$$

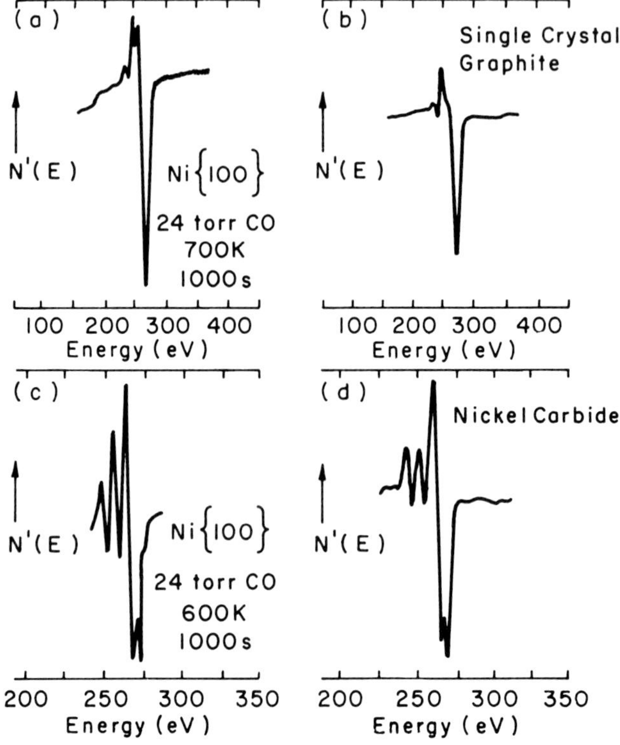

FIGURE 5. Comparison of AES carbon singles on a Ni(100) crystal with those single crystal graphite and nickel carbide. (From Goodman, D. W., Kelley, R. D., Madey, T. E., and Yates, J. T., *J. Catal.*, 63, 226, 1980. With permission.)

$$CH_2(a) + H(a) \rightarrow CH_3(a) \quad (11)$$

$$CH_3(a) + H(a) \rightarrow CH_4 \quad (12)$$

The oxygen produced by CO dissociation is removed from the catalyst by the reaction of atomic hydrogen as H_2O. This reaction is very fast as compared with methanation. CO_2 is also produced, but only at a level of about 2% of the CH_4.[89]

The nature of the controlling step is not yet conclusively established. Nevertheless, based upon the turnover numbers (the turnover number for carbide removal at 450 K was 2×10^{-4} compared to the carbide formation, 2×10^{-4}, and methane formation, 3×10^{-4}), Goodman et al.[27,28] concluded that the rate of methanation over single crystals of nickel is determined by a subtle balance of the carbide formation and removal steps. Goodman's group[88] also obtained very similar results for single crystals of ruthenium.[88]

More recently, Reaction 12 has been studied by Yates et al.[90] These authors used the thermal decomposition of ethanol on Ni(111) to provide an efficient source of $CH_3(a)$, and concluded that the rate-limiting step(s) in methane synthesis from CO and H_2 exist at an earlier stage than the step 12 elementary process.

It appears that methanation over both Ni and Ru, as well as in general over other transition metals (Fe and Co), follows the same reaction mechanism since it shows extraordinary similarities in so many critical parameters.[24] Direct evidence for elementary step 9 on Ru(001),[171,172] and for steps 9 to 11 on Ni(111)[173] is now available. These species were primarily identified by HREELS[171] and SIMS.[172,173]

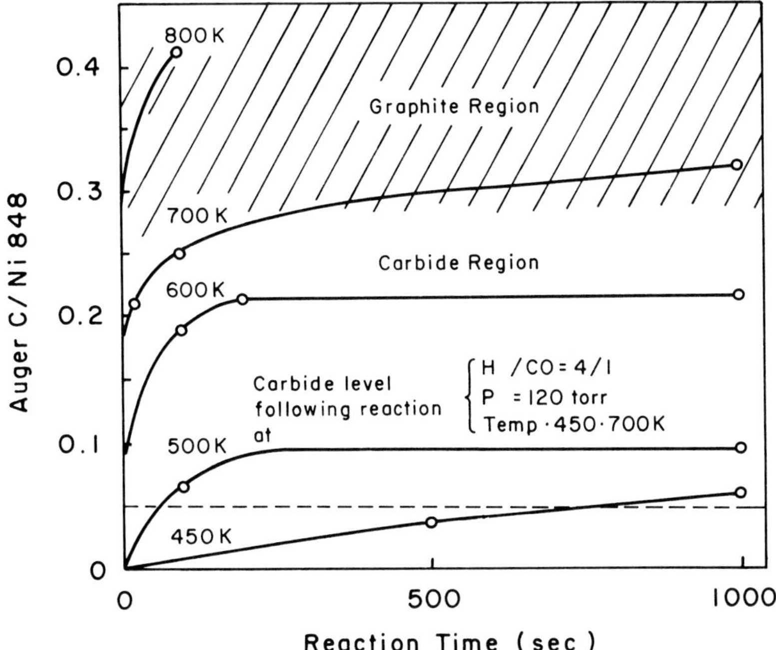

FIGURE 6. The rate of carbon build-up on Ni(100) catalyst by reaction with CO. Ordinate values of carbon Auger intensity normalized to the Ni 848 eV transition Auger intensity. (From Goodman, D. W., *J. Vac. Sci. Technol.*, 20, 522, 1982. With permission.)

In conclusion, the mechanism involves submonolayer concentrations of an active, carbidic surface species under reaction conditions. The rate-determining step possibly involves steps 7, 9, 10, or 11, or perhaps a combination of steps. It is anticipated that future studies with the UHV-high-pressure techniques will help to achieve an even more detailed picture of the reaction mechanism.

G. Catalyst Poisoning by Sulfur

Methanation catalysts are very sensitive to the low concentrations of sulfur-containing compounds which are inevitably present in the feed. The regeneration of sulfur-poisoned Ni catalysts has not been successfully demonstrated.[92,93] The precise action of sulfur on nickel is uncertain. Two general reviews on sulfur adsorption and poisoning of metallic catalysts have appeared recently.[94,108]

Sulfur adsorption on metal surfaces may alter their adsorption characteristics for various reactant molecules, either by blocking active surface sites and thus preempting reactant molecules (geometric effects), or by electron density changes caused by strong metal-sulfur interactions (electronic effects). Therefore, the most important question has been whether sulfur poisoning is due to geometric effects or whether electronic effects are playing a major role in the deactivation process.

In recent years, these fundamental questions have been attempted essentially in two ways. The first line of investigation pursued, in an effort to characterize sulfur poisoning, is the kinetic study of the reaction and the effect sulfur has on methanation activity. The second is the utility of surface techniques in studying sulfur poisoning on Ni-Al_2O_3 as well as with single crystals of nickel.

Reaction studies have shown that the activity of Ni-Al_2O_3 dropped by 2 to 3 orders of magnitude in the temperature range 525 to 660 K in as little as 13 ppb H_2S.[67,95,96] Fitzharris

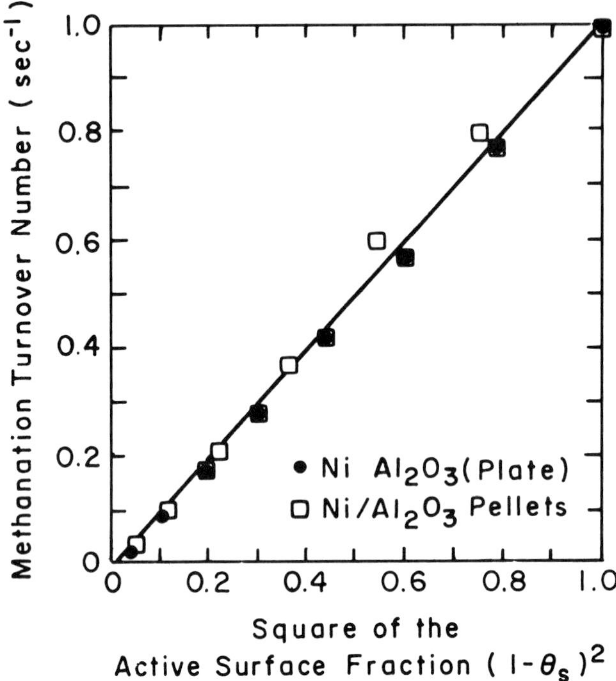

FIGURE 7. Variation in methanation rate of Ni/Al_2O_3 with the square of the unpoisoned surface fraction. Reaction conditions: 660 K, p_{H2} = 1 atm, p_{CO} = 0.044 atm, p_{H2S} = 55 ppb. (From Fitzharris, W. D., Katzer, J. R. and Monogue, W. H., *J. Catal.*, 76, 369, 1982. With permission.)

et al.[95] have shown that the methanation rate, relative to the fresh catalyst, is directly proportional to the square of the fraction of the unpoisoned surface (Figure 7). Their study also showed that methanation activation energy is identical for fresh and completely poisoned catalyst (100 kJ/mol). Rostrup-Nielsen and Pederson[96] also reported strong nonlinear deactivation, but the activation energy remained constant. Fitzharris et al.[95] concluded that, (1) poisoning of Ni by sulfur is primarily a geometric effect in which sulfur blocks the active sites making them inaccessible for reaction and (2) the rate-controlling step in methanation requires two surface sites.

Surface analysis techniques continue to be useful in active site identification. It has been shown by Ng and Martin,[97] by means of magnetic measurements of nickel films, that H_2, like CO, is not adsorbed on a surface which is completely poisoned by sulfur. Rewick and Wise[98] studied CO adsorption over sulfur-poisoned Ni/Al_2O_3 with IR spectroscopy. It has been concluded that in the presence of H_2S, the adsorption sites for the bridged bond (high-coordination species) CO was greatly reduced, and to that relatively high coverages of sulfur, the linear bonding sites were unaffected.[98,99]

During the initial stages of sulfur adsorption on clean nickel single-crystal faces, sulfur atoms reside in high-coordination sites. Earlier studies,[100,101] using LEED, photoemission, and other techniques, showed that, at low coverages (one quarter monolayer), the sulfur is adsorbed on a Ni (100) plane as an ordered p (2 × 2) overlayer with each sulfur atom bonded to 4 Ni atoms. With increasing coverage (one half monolayer) the structure changes to an ordered c (2 × 2) pattern; the one half monolayer corresponds to a saturation coverage.

More recent studies (UHV-high pressure) of Goodman and co-workers[24,27,28,102,105] have shown that the dominant effect in poisoning of the methanation reaction over Ni(100) and Ni(111) by sulfur is an electronic one and extends over distances larger than the atomic

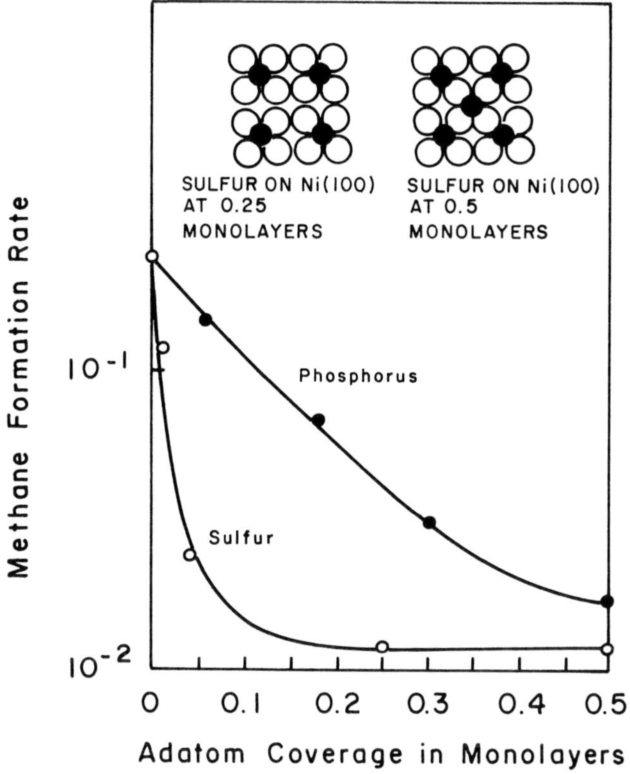

FIGURE 8. Methanation rate as a function of sulfur and phosphorus coverage on a Ni(100) catalyst. Reprinted with permission from Goodman, D. W., *Acc. Chem. Res.*, 17, 194, Copyright (1984), American Chemical Society.

radius. LEED patterns were used for surface coverage determination by AES by providing calibration points. AES was then utilized to probe the CO-molecular and metal bonds in the presence of sulfur.[104]

Both the kinetics and TPD study[102] showed that the poisoning effect of sulfur was very nonlinear (Figure 8) and was identical to those observed for supported Ni-Al_2O_3.[96] Apparently, sulfur bonded in the p (2 × 2) configuration sufficiently deactivates the catalyst surface for methanation, and further addition of sulfur produces no reaction rate attenuation. The unoccupied fourfold nickel sites remaining at a sulfur coverage of 0.25 monolayer are effectively poisoned for carbon formation and carbon hydrogenation catalysis. A rapid initial drop in the rate at low sulfur coverages in Figure 8 led these authors to conclude that ten or more equivalent nickel sites are deactivated by one sulfur atom.

In order to show that the main sulfur-poisoning mechanism is electronic, Kiskanova and Goodman[103] substituted phosphorus for sulfur, since then the reaction rate will be expected to be a function of the relative electronegativity of the poison. They argued that if a ten-nickel atom ensemble is required for methanation, then changing the electronic character of the poison should produce little change in the reaction rate; the requirement for an ensemble effect is that a certain number of surface atoms are necessary for a reaction to occur. Indeed, they observed a distinct change by replacing sulfur with phosphorus, because of its less electronegative character, as illustrated in Figure 8. Chemisorption of H_2 and CO on a Cl, S, and P (atomic and covalent radii are similar, 0.99, 1.04, 1.10 Å, respectively) partially covered Ni (100) surface support corroborates this conclusion.[24,102,103,106] Their contention of long-range effects was further supported by a similar observation of nonlinear poisoning of nickel by sulfur for CO_2 methanation.[107]

In contrast, Madix et al.[177] claimed that the effect of sulfur is local and not a long-range electronic effect. These authors argued that the desorption state characteristic of clean Ni(100), shown to be eliminated at sulfur coverage of 0.25 monolayer in the studies of Goodman and Kiskanova,[103] alone is insufficient to show long-range interactions between CO and sulfur, since in the p (2×2)S structure each nickel atop site is easily blocked by the sulfur atom in the fourfold hollow. Their desorption results[177] further showed that sulfur converts the high temperature state into two states at low temperatures. This led them to affirm that a small amount of sulfur forced CO off the atop binding sites at saturation, thereby altering the desorption behavior, and thus contradicting with long-range interactions.

In summary, the balance sheet of the studies conducted on crystallographically and chemically well-defined surfaces can be considered very positive for the understanding of the mechanism responsible for the poisoning of nickel catalysts by sulfur. However, the conclusion that poisoning by sulfur at low coverages is long range rests on more subtle considerations.

IV. METHANOL SYNTHESIS

A. Reactions, Catalyst, and the Process

The synthesis gas preparation phase of methanol synthesis is presently the most in need of major changes and developments. It is about the only route from synthesis gas to a single organic compound which can be made in high yield at relatively low cost with the present technology. The potential of methanol for chemical feedstock and other uses has been widely recognized.

The plant essentially consists of a steam reformer and the methanol converter. Steam-hydrocarbon reforming has become the most frequently used method for synthesis gas preparation, and usually natural gas (mostly methane) is converted into mainly CO, CO_2, and H_2 (supported nickel oxide).[109-112] The basic equation is

$$CH_4 + H_2O = 3H_2 + CO \tag{13}$$

However, for methanol synthesis, the following reaction is necessary:

$$CO + 2H_2 = CH_3OH \tag{14}$$

Therefore, the following simultaneous operation is performed in the reformer by adding CO_2 in the feed gas:

$$CO_2 + CH_4 \rightleftharpoons 2CO + 2H_2 \tag{15}$$

Equation 15 produces excess CO relative to that needed for methanol, and as a result, in commercial practice, the feed to a reformer comprises natural gas, steam, and carbon dioxide.

The synthesis gas mixture which enters the methanol converter typically consists of around 10% CO and a few % CO_2, the balance being H_2.[110] The other major reaction involved is

$$CO_2 + 3H_2 = CH_3OH + H_2O \tag{16}$$

Both Reactions 14 and 16 are exothermic, the reaction heat being −91 and −50 kJ/mol, respectively. Under the synthesis conditions, the water-gas shift reaction can also take place according to Reaction 2. It has been suggested that, if methanol synthesis proceeded through only one of the routes, 14 or 16, then Reaction 2 could form a route by which the concentration of either CO or CO_2 could be maintained at a high level.[110]

Current technology, pioneered by ICI® in 1965, and later followed by other companies, requires relatively high pressure (50 to 100 atm) and temperature (520 to 550 K) operating conditions in the conversion of synthesis gas.[110-111] At these temperatures, which need to be used in order to obtain sufficient catalytic activity, there is an equilibrium limitation that results in 6% of methanol in the reactor effluent. Compression and recycle costs are high. To circumvent these difficulties and to lower the cost of methanol production, lower temperatures and lower pressures of operation are being considered. In addition, the Cu-Zn-based catalysts are very sensitive to sulfur poisoning, and therefore, the current technology demands a thoroughly desulfurized feed stream of reactants.

Ternary compositions Cu-ZnO-Al_2O_3 and Cu-ZnO-Cr_2O_3 are the most important industrial methanol synthesis catalysts. Typically, the ICI® catalysts consist of Cu-ZnO-Al_2O_3 in the metal atomic ratio of approximately 60:30:10.[110] In actual practice, these catalysts contain a mixture of microcrystalline copper, microcrystalline zinc aluminate, and most likely zinc oxide.[113] Andrew[114] emphasized the importance of $ZnAl_2O_4$ in stabilizing the catalyst for long-term performance.

There are other components such as alkali, other transition metals, and rare earths which have been used in combination with copper and zinc oxide. For example, Rb or Cs is very effective in reducing methanation without sacrificing too much alcohol productivity. Recently, a quarternary catalyst has been patented[115] which incorporated the components of both the methanol catalysts as well as of the Fischer-Tropsch catalysts. The catalyst is claimed to be effective in producing a mixture of C_1-C_4 alcohols from the synthesis gas. There is yet little study on the mechanistic details.

The production of methanol from CO and H_2 over other metallic catalysts[116-124] of Pd, Pt, Ir, and Rh, as well as homogeneous catalysts,[125-128] has been reported recently. While these are of great interest, emphasis will be placed on findings related to the current commercial Zn-Cu-based catalysts.

B. Interaction of CO-CO_2-H_2 Over Zn-Cu System: An Overview

There has been much discussion recently about the mechanism of synthesis on Zn-Cu catalysts, and many aspects of the mechanism are still not fully understood.[112,113] The controversy exists about the nature of active sites as well as the structure of some surface intermediates.

Kung,[112] in his 1980 review, discussed a successive hydrogenation of CO as well as of reaction procedure based on carbon monoxide with a formate intermediate stage which is subsequently hydrogenated and dehydrogenated. In contrast to the conversion of CO into methanol, the conversion of CO_2 into methanol involves a pathway, in which the oxygen atom of the formate group is attached to the metal surface which is known to have a lower activation energy.[129] Russian researchers,[130-132] however, favored the formation of methanol on Cu-ZnO catalysts by hydrogenation of CO_2 which is formed during CO-water-gas reaction. They did not observe any methanol formation when the synthesis gas was free of CO_2 and H_2O.

Recent studies in several laboratories clearly show that carbon dioxide has a significant effect on the synthesis rates[113,131,133,134] and this rate is maximum around 1% CO, a result derived from tests involving the commercial ICI® catalyst (Cu-ZnO-Al_2O_3).[114] Under industrial conditions, more recently, ICI researchers[135] have further shown by using labeled carbon oxides that all of methanol is formed from the CO_2 component of the feed rather than the CO. The major reactions are

$$CO_2 + 2H_2 \rightarrow CH_3OH + O(a) \qquad (17)$$

$$CO + O(a) \rightarrow CO_2 \tag{18}$$

$$H_2 + O(a) \rightarrow H_2O \tag{19}$$

where $O_{(a)}$ is a surface oxygen atom. Therefore, the oxidation state of the catalyst surface will be controlled by the relative rates of Reaction 17, 18, and 19. As a result, the copper surface will be covered with formate, formyl, methoxy species, and oxygen to an extent determined by the kinetics and relative rates of the various reaction steps.[136]

There is general agreement that copper is the active component, and the catalyst activity is proportional to the copper surface area.[114,135] However, a considerable amount of discrepancy exists in the assignment of the chemical state of the copper component in the catalyst. According to Herman and co-workers,[113,137-139] a Cu^+ solution in ZnO is an active phase in which Cu^+ nondissociatively chemisorbs and activates Co whereas ZnO activates the hydrogen. It was also established that neither chromia nor alumina were essential for the high catalyst activity and selectivity. They further claimed that reduced copper itself has zero activity for methanol synthesis within detection limits. In contrast, Andrew,[114] who from a study of methanol and water-gas shift catalysts, presented a case for copper being the only active component and the role of ZnO was to adsorb poisons present in the synthesis gas. Studies on Raney® Cu-Zn[140,141] catalysts have also shown that the active component is the metallic Cu.

From this brief review, it is apparent that the mechanistic details of the catalysts and reactions involved in the synthesis are rather complex. It must, however, be recognized that unlike methanation studies, the UHV studies of methanol synthesis from synthesis gas components are likely to be unsuccessful because of the very low maximum yields of methanol at such low pressure.[110] This is the reason why UHV works mainly concentrated on the decomposition of methanol rather than on its synthesis.

The following sections will therefore attempt to present the results by separating out different components of this synthesis in order to gain a basis from which to progress to the overall function of the catalyst. However, before they are introduced, the kinetic equations are outlined in the context of process design.

C. Kinetic Models

The early studies were made over 30 years ago over zinc chromite catalysts, and the derived kinetic equation did not contain a CO_2-dependent term at all. Using data obtained over the widest range of pressure, ratio of H_2/CO, and temperatures 570 to 630 K, Natta et al.[144] obtained a good fit of data with the expression for the methanol formation rate r, where a trimolecular reaction takes place on the surface between a CO molecule and two adsorbed H_2 molecules:

$$r = \frac{(\gamma_{H_2}^2 \cdot \gamma_{CO} - \gamma_{CH_3OH})/K_p}{(A + B\gamma_{CO} + C\gamma_{H_2} + D\gamma_{CH_3OH})^3} \tag{20}$$

where $\gamma_i = f_iP_i$ is the fugacity coefficient of species i; P_i is the partial pressure of species i; K_p is the equilibrium constant of carbon monoxide hydrogenation to methanol; and A, B, C, and D are empirical constants.

Another rate expression was proposed by Uchida and Ogino[145] from high-pressure data, assuming that the desorption of methanol was the rate-controlling step:

$$r = k(p_{H_2}^2\, p_{CO})^{0.7} \tag{21}$$

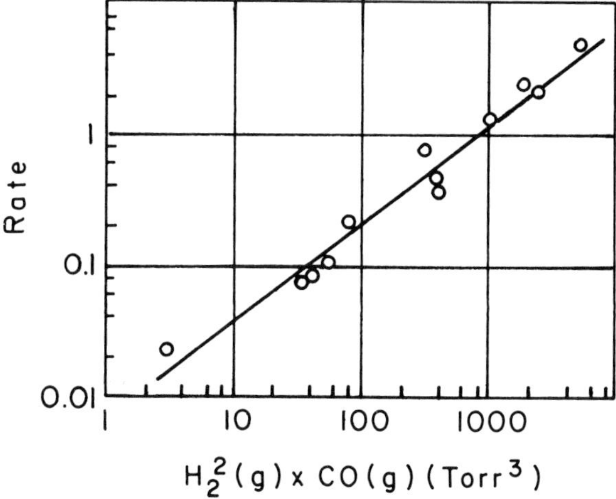

FIGURE 9. Methanol synthesis rate as function of the product of concentrations $[H_2]^2 \times [CO]$ over a $Cu/ZnO/Cr_2O_3$ catalyst at 523 K.[148]

The activation energy was reported to be 108 kJ/mol. Similar to early kinetic studies, Leonov et al.,[146] in 1973, proposed a kinetic rate equation which did not contain the CO_2-dependent terms.

The classical rate expression of Natta, which contains four empirical constants, had been altered or extended by other authors to fit a wider range of data, which were summarized by Denny and Whan[133] up to 1977. With the realization that CO_2-dependent terms had to be incorporated in order for the kinetic equation to be more realistic, Bakermeir et al.[147] obtained a rate expression for the ZnO/Cr_2O_3 catalyst. The resulting equation, however, contained carbon dioxide as a retardant rather than as a reactant or a promoter.

Andrew,[114] in 1980, presented a kinetic expression that includes an empirical CO_2-dependent term for the copper-zinc oxide-alumina of the form:

$$r = k\, p_{CO}^{0.2-0.6}\, p_{H_2}^{0.7}\, \Phi_{CO_2} \tag{22}$$

It was reported that the rate of methanol synthesis reached a maximum at the CO_2 to CO partial pressure ratio around 0.01 to 0.03 and decreased as the CO_2 pressure was further increased, though the function Φ was not distinctly ascertained. It was also observed that the synthesis rate would decline at very small concentrations of CO_2. A similar behavior was obtained in the binary Cu-ZnO catalysts.[137]

So far, there are three features common to all kinetic observations. First, there is a critical CO_2/CO ratio, below which the reaction does not proceed. Second, an optimum concentration of CO_2 exists at which the synthesis rate is maximum. The third common feature is that the synthesis rate depends on positive powers of partial pressure of hydrogen and carbon monoxide, the dependence of H_2 being stronger than that of CO, such as illustrated in Figure 9 for the Cu-ZnO-Cr_2O_3 catalyst.[148]

To account for these observations, Klier et al.,[149] in 1982, presented a model by carrying out a detailed study of the kinetics of methanol synthesis over copper-zinc oxide catalysts at a pressure of 75 atm, temperatures of 498, 508, and 523 K, and with varying $CO_2/CO/H_2$ ratios. The experimental results were analyzed on the basis of redox kinetics with statistical data interpretation. They formulated the model for the synthesis as follows:

1. The catalyst can exist in a reduced state Cat· and in an oxidized state Cat–O. The oxidized state is active. The proportion of Cat–O and Cat· is controlled by the ratio of CO_2 and CO in the synthesis gas.
2. Several active centers Cat–O are involved in each reaction step. These centers may be identified with copper solute species in zinc oxide.
3. All three components of the synthesis gas, CO, H_2, and CO_2 react in the adsorbed layer; CO_2 competes for active sites with at least one of the reactants CO and H_2. The adsorption strengths are in the order $CO_2 > CO > H_2$.

The redox equilibrium that controls the active form Cat–O is

$$\text{Cat–O} + \text{CO(g)} \rightleftharpoons \text{CO}_2\text{(g)} + \text{Cat}\cdot \tag{23}$$

The data were most satisfactorily correlated by the general form which may be considered as a refined version of the original Natta kinetics:[113,149]

$$r = \text{const } (\text{Cat-O})^m \frac{p_{CO}\, p_{H_2}^2 - p_{MeOH}/K_{eq}}{(F + K_{CO_2}\, p_{CO_2})^n} + k'[p_{CO_2} - (K'_{eq})^{-1}(p_{MeOH}\, p_{H_2O}/p_{H_2}^3)] \tag{24}$$

where F is the linear function of pressures of H_2, CO, and H_2O, and is determined by the ratio P_{CO_2}/P_{CO}; and, K_{eq} and K'_{eq} are for Reactions 14 and 16, respectively. The comparison of the experimental points with the theoretical kinetic curves (integrated form of Equation 24) with m = 3, and n = 3 is shown in Figure 10.

Although Klier et al.[113,149] determined the kinetics for the binary Cu/ZnO catalysts, their rate expression should be able to fit a wider range of data obtained with commercial catalysts. In principle, the CO_2/CO dependence of the synthesis rates over the ICI® catalysts can be described by the same formal kinetics, since both sets of data require the reduced form of the catalyst to be inactive, despite the disagreement on the chemical nature of Cat· at low temperatures.[113]

In summary, it appears likely that the kinetic expression for methanol synthesis is governed by the same set of solid-state reactions underlying the CO_2/CO dependencies over the binary Cu/ZnO and the ternary Cu/ZnAl$_2$O$_4$/ZnO catalysts, irrespective of the chemical nature of the redox process involved in the formation of Cat–O.

D. Mechanism

1. The Copper Component

Copper is a weakly adsorbing metal except for the chalcogens (O,S); the heat of adsorption of CO is only 55 kJ/mol, and CO_2 as well as H_2 is probably even less.[150-152] Balooch et al.[152] made the molecular beam measurements and concluded that hydrogen adsorption is activated with the initial sticking coefficient of less than 10^{-2} even at high substrate and gas temperatures.

The activity for methanol formation of chromia-supported copper has been shown to be related to the amount of surface-stable Cu^+ detected by XPS.[166] These authors claimed to have reported the first direct evidence for Cu^+ as the associative CO (i.e., alcohol producing) on copper chromia catalysts. They also found a surface concentration of Cu^+ stable to hydrogen reduction at 543 K and correlated the surface concentration of Cu^+ with methanol synthesis activity (for a synthesis gas of 67% H_2/33% CO).

In contrast, Chinchen et al.[136] have found that the surface of copper supported on a variety of oxides (Al_2O_3, SiO_2, MgO, and ZnO) contains both metallic and oxidized copper under

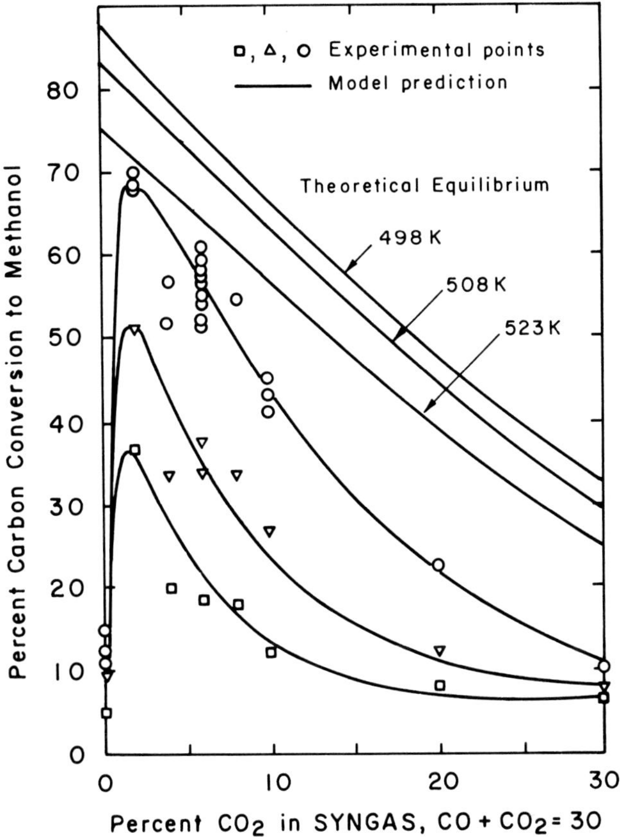

FIGURE 10. The dependence of methanol synthesis rates at 498 to 523 K, 75 atm total pressure upon the ratio of concentrations of CO_2 and CO. The total hydrogen to carbon ratio in the feed gas was 7:3 for all CO_2/CO ratios. (From Klier, K., Chatikavanij, V., Herman, R. G., and Simmons, G. W., *J. Catal.*, 74, 343, 1982. With permission.)

real methanol synthesis conditions; the greater the CO_2/CO ratio the more oxidized is the copper surface. In comparison with supported copper-based catalysts, Herman and coworkers[137] showed that copper metal is a very poor catalyst for methanol synthesis at the synthesis conditions.

The very recent work on the adsorption of methanol has clarified the chemical state of the adsorbate on single-crystal copper surfaces. Methanol is known to adsorb associatively on copper at low temperatures whereas dissociation occurs at room temperature.[153-154] Bowker and Madix[153] used UPS, XPS, and TDS to identify products of methanol adsorption on the Cu(110) crystal face and determined a half-monolayer coverage of methoxy species originating from methanol dissociation on this surface at 270 K.

SIMS has also been used successfully for inferring the chemical state of methanol on Cu(100).[155] Physisorbed methanol molecules present at 120 K were found to be readily differentiable from the methoxy species formed at 330 K. The SIMS spectra of methanol adsorbed on Cu(100) at 120 and 300 K are shown in Figure 11.[155] For the adsorption at 300 K (Figure 11A) characteristic peaks of the adsorbate include: $CuOCH_n^+$, n = 2,3; $CuCH_3^+$; and $Cu_2CH_3^+$. The absence of $Cu(CH_3OH)^+$ which is, however, a stable-ion species, and the presence of $(CuOCH_3)^+$ reflect significantly that methoxy species are chemisorbed at room temperature. The abundance of $(CuOCH_2)^+$ ions has been attributed by Marien et

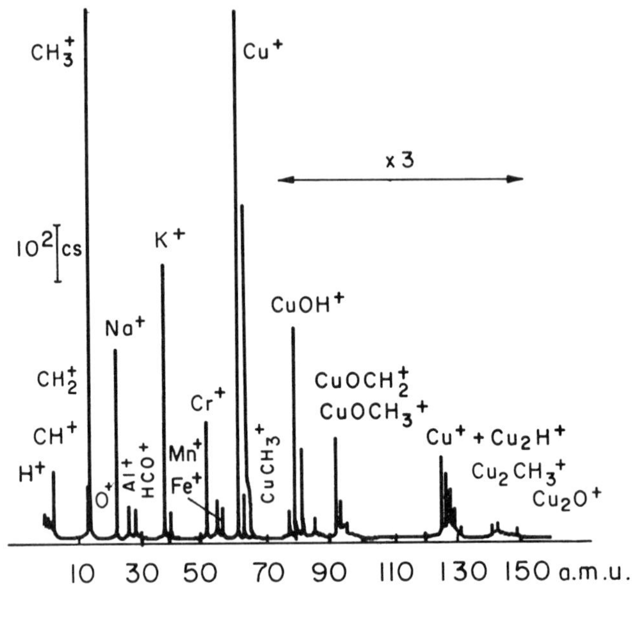

A

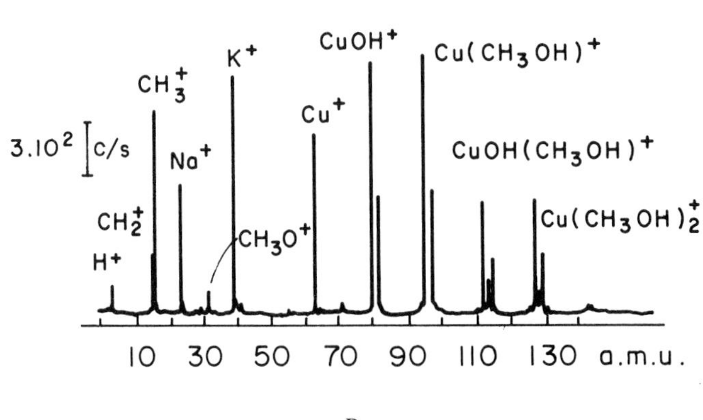

B

FIGURE 11. Positive static SIMS spectrum of Cu(100) covered by (A) 6L of CH_3OH at 300 K and (B) 1L of CH_3OH at 120 K. (Reprinted with permission from Marien J., Depauw, E., and Peizer, G., *J. Phys. Chem.*, 87, 4344, Copyright (1983) American Chemical Society.

al.[155] to a fragmentation of most of the emitted $(CuOCH_3)^+$ parent ion based on their HREELS work which shows the absence of C–H softening for $CH_3O_{(a)}$.

Adsorption at 120 K, however, produces strong cluster ions containing the methanol molecule — $Cu(CH_3OH)_n^+$ and $CuOH(CH_3OH)_n$, n = 0, 1, 2. This is illustrated in Figure 11B, where the intensity of Cu^+ is still high. In summary, the absence of the $CuOCH_3^+$ cluster ion at 120 K reflects the fact that methanol is molecularly adsorbed on Cu at this temperature and is dissociatively chemisorbed at 300 K.

2. *The Zinc Oxide Component*

Like metallic copper, pure zinc oxide is a poor methanol catalyst at synthesis conditions. In contrast, it has relatively high activity at pressures exceeding 200 atm and temperatures above 623 K.[113] There is general agreement that a ZnO surface is present during methanol synthesis.[113,143]

It is known that specific activity of zinc oxide in methanol synthesis increases with increasing surface area.[156] Although the resulting catalysts were poorly characterized by current standards, this behavior could be the result of these high-area zinc oxides subtending a greater portion of a more active crystal plane; a number of studies with single crystals of ZnO have demonstrated very different properties on different crystal surfaces.[157,158]

Zinc oxide crystallites have a hexagonal wurtzite-type structure. Much of the work has been carried out on the natural nonpolar $(10\bar{1}0)$ and $(11\bar{2}0)$ planes, and the polar (0001) and $(000\bar{1})$ planes. A perfect surface of nonpolar $(10\bar{1}0)$ has an equal number of anions and cations in the same plane, and is very stable.[158-160] On the other hand, a perfect polar (0001) would have a layer of zinc ions that is outwardly situated with respect to the next layer of oxygen atoms, and a perfect $(000\bar{1})$ surface would have a layer of oxygen ions farther out than the next layer of zinc ions.[112]

A very important observation regarding the activity of zinc oxide crystal faces for the hydrogenating properties, originally suggested by Grunze et al.[161] and later confirmed by Bowker et al.,[157] is that only crystals of zinc oxide exhibiting a reasonable percentage of polar faces showed pronounced Zn–H and O–H bands in the infrared spectrum after exposure to hydrogen. These findings led Bowker et al.[157] to conclude that the adsorption of hydrogen and carbon dioxide, and their surface reaction to form formate and then methanation all take place on the polar faces of the zinc oxide only. More recent single-crystal studies further confirmed that the zinc in polar face is indeed chemically active, while the O-polar face is inactive.[174-176]

There have been several studies on the adsorption and interactions of hydrogen, water, carbon monoxide, carbon dioxide, formaldehyde, and methanol with zinc oxide. Surface formate and methoxide bands have been detected by IR in methanol decomposition over zinc oxide.[162] These authors have further shown that surface formates are formed on ZnO, from CO_2 and H_2 but not from CO–H_2 or CO–H_2O. The formation took place at temperatures above 453 K, which decomposed into CO and hydrogen at 573 K at the ratio of CO to H ≡ 1:1.

Bowker et al.[163] proposed a mechanism wherein carbon oxides and H_2 react via a formate to methanol over zinc oxide. This was based on the desorption spectra obtained for methanol adsorption on the prism face of zinc oxide which were virtually identical to those obtained from formaldehyde chemisorption. The same formate intermediate was formed by coadsorption of CO_2 and H_2 while H_2/CO did not result in its formation. The role of the CO was attributed to be that of a reducing agent, maintaining the zinc oxide in a sufficiently anion-deficient state. No microreactor experiments were conducted on clean zinc oxide surface.

3. *The Combined Cu/ZnO System*

Although the bifunctional nature of the catalyst is different, some inferences will be drawn from the above discussion. All the binary Cu/ZnO catalysts were definitely highly selective toward methanol than either simple copper catalysts or zinc oxide catalysts, the latter only having good activity under high pressure and temperature conditions.

As presented earlier, there is still controversy about the chemical state of copper component both in binary and in ternary catalysts during methanol synthesis. Herman and coworkers[137-139,149,164] combined reactivity studies with bulk techniques (diffuse reflectance spectroscopy, STEM, and XRD) to correlate the active phase of CuO-ZnO methanol synthesis catalysts with Cu^+ dissolved in the ZnO lattice. In two recent papers,[165(a,b)] precursor, calcined, and reduced methanol catalysts were surface characterized by ESCA. Okamoto et al.[165] proposed that both Cu^+ and an electronically altered Cu^o are the active sites. These authors, however, assumed that the Cu^+ species is directly responsible for the reaction as claimed by Herman et al.

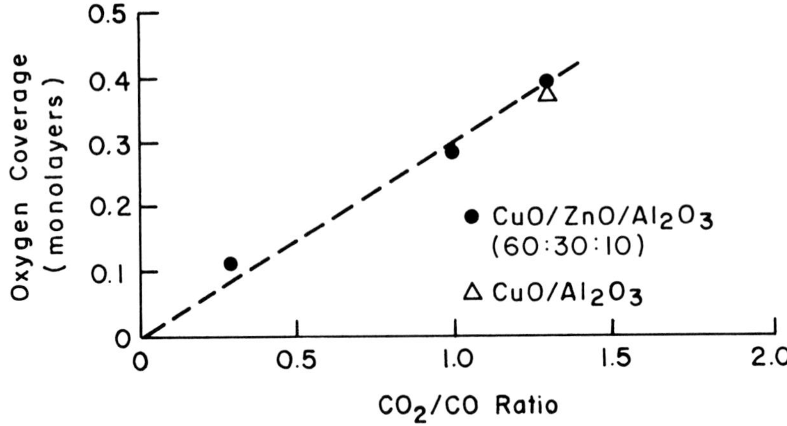

FIGURE 12. Steady-state oxygen coverage of the copper surface under methanol synthesis conditions as a function of the CO_2/CO ratio. (From Chinchen, G. C., Waugh, K. C., and Whan, D. A., *Appl. Catal.*, 25, 101, 1986. With permission.)

Research has also been carried out to elucidate the chemical state of the copper in the commercial catalyst surface (CuO-ZnO-Al_2O_3). The results are, however, contradictory to the above proposal. In the two very latest papers, ICI researchers[136,143] reported that the surface of CuO-ZnO-Al_2O_3 contains both metallic and oxidized copper under real methanol synthesis conditions and the proportion of oxidized surface depends on the exact composition of the synthesis gas. These authors have shown that the steady-state oxygen coverage of the copper surface under methanol synthesis conditions for both Cu/ZnO/Al_2O_3 and Cu/Al_2O_3 is essentially a linear function of the CO_2/CO ratio (Figure 12); the greater the CO_2/CO ratio the more oxidized is the copper surface.

Their contention is further supported, if not fully but at least in part, by a high pressure-UHV study on the surface of a commercial CuO-ZnO-Al_2O_3. Fleisch and Mieville[142] showed that the surface was entirely metallic copper, ZnO and Al_2O_3. However, their synthesis gas composition (73% H_2, 25% CO, and 2% CO_2) was rather unusual, since it had a CO_2/CO ratio of only 0.08. This would result in an extremely low oxygen coverage (<2%), and probably will be undetected by their XPS technique.

ICI researchers[136] have further reported the methanol synthesis activity on the copper metal area under methanol synthesis conditions (50 atm, 510 K) for a variety of catalysts (Figure 13). These authors concluded that all the copper is active and only the copper surface is implicated in the rate-determining step of the synthesis.

4. Reaction Pathways

There are relatively few definitive mechanism studies on the methanol synthesis reactions. UHV studies have been unsuccessful due to unfavorable pressure conditions. As a result, the kinetically significant intermediates are difficult to detect during the synthesis, and therefore, the synthesis mechanism remains unclear.

Klier[113] has summarized the mechanisms proposed in the literature into three basic groups:

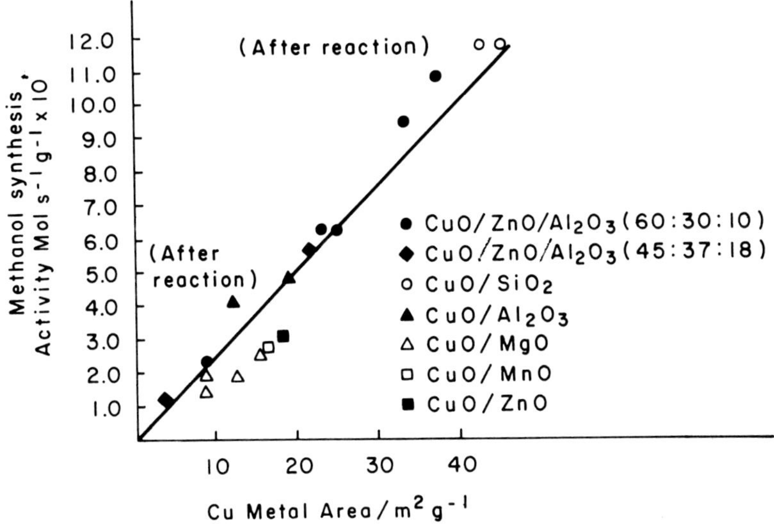

FIGURE 13. Methanol synthesis activity as a function of copper metal area. From Chinchen, G. C., Waugh, K. C., and Whan, D. A., *Appl. Catal.*, 25, 101, 1986. With permission.)

$$CO \rightarrow \underset{\underset{M}{\|}}{\overset{\overset{O}{\|}}{C}} \xrightarrow{H\text{-}M} \underset{\underset{M}{|}}{\overset{H \diagdown \quad /\!\!/ O}{C}} \xrightarrow{H\text{-}M} \underset{\underset{M}{\|}}{\overset{H \diagdown \qquad \diagup H}{C\text{–}O}}$$

$$\xrightarrow{H\text{-}M} H\text{–}\underset{\underset{M}{|}}{\overset{\overset{\overset{H}{/}}{\overset{O}{|}}}{C}}\text{–}H \xrightarrow{H\text{-}M} CH_3OH + M \qquad \text{(I)}$$

$$\underset{\underset{M}{|}}{\overset{\overset{H}{|}}{\ddot{O}}} \xrightarrow{CO} \underset{O \;\; M \;\; O}{\overset{\overset{H}{|}}{C}} \xrightarrow{4\,|^{M}_{H}} \underset{\underset{M}{|}}{\overset{\overset{CH_3}{|}}{O}} + H_2O \rightarrow \underset{\underset{M}{|}}{\overset{\overset{H}{|}}{O}} + CH_3OH$$

$$\downarrow H_2$$

$$\underset{\underset{M}{|}}{H} + CH_3OH \qquad \text{(II)}$$

$$\mathrm{CO} \rightarrow \begin{array}{c}\mathrm{O}\\ \|\\ \mathrm{C}\\ \|\\ \mathrm{M}\end{array} \xrightarrow{\begin{array}{c}\mathrm{H}\\ 2|\\ \mathrm{M}\end{array}} \begin{array}{c}\mathrm{H}\diagdown\\ \mathrm{C{=}O}\\ \mathrm{H}\diagup\ \ |\\ \mathrm{M}\end{array} \xrightarrow{\begin{array}{c}\mathrm{H}\\ 2|\\ \mathrm{M}\end{array}} \mathrm{M} + \mathrm{CH_3OH} \qquad \text{(III)}$$

In mechanism I the reaction proceeds by successive hydrogenation of CO to form the corresponding intermediates. In pathway II the first step of the reaction is the insertion of CO into a surface hydroxyl to form a surface formate. While the intermediates in I are bonded to the surface cation via the carbon atom, in pathway II the bonding is via the oxygen atom. Mechanism III requires formaldehyde as an intermediate. While formaldehyde has never been observed as an intermediate in methanol synthesis over catalysts reviewed here, it has been reported to be the product of methanol decomposition.[153,154]

To reconcile with the marked differences in the ease of transformation of carbon dioxide and carbon monoxide into methanol, together with increased selectivity associated with the CO_2 reaction, Denise et al.[167,168] proposed two routes to methanol. The first, from CO, where the carbon atom of the intermediate is attached to the metal surface:

$$\mathrm{CO} \xrightarrow{\mathrm{M}} \mathrm{MCO} \xrightarrow{\mathrm{H_2}} \mathrm{M{-}\overset{\overset{\displaystyle \mathrm{H}}{|}}{C}HO} \xrightarrow{\mathrm{H_2}} \mathrm{M{-}\overset{\overset{\displaystyle \mathrm{H}}{|}}{C}H_2OH} \rightarrow \mathrm{CH_3OH}$$

$$\mathrm{M{-}\overset{\overset{\displaystyle \mathrm{H}}{|}}{C}HO} \rightleftharpoons \mathrm{M{=}C}\begin{array}{l}\diagup \mathrm{H}\\ \diagdown \mathrm{OH}\end{array} \qquad \text{(IV)}$$

The second, from CO_2, proceeds via O-bonded entities, produced by the reduction of carbonates, formates, etc.:

$$\mathrm{CO_2} \xrightarrow{\mathrm{M{-}OH}} \mathrm{M}\begin{array}{c}\mathrm{O}\\ \diagup\ \ \diagdown\end{array}\mathrm{C{-}OH}\ (\mathrm{C}{=}\mathrm{O})$$

$$\downarrow$$

$$\mathrm{CO_2} \xrightarrow{\mathrm{MH}} \mathrm{M}\begin{array}{c}\mathrm{O}\\ \diagup\ \ \diagdown\end{array}\mathrm{C{-}H}\ (\mathrm{C}{=}\mathrm{O}) \xrightarrow{\mathrm{M'}} \begin{array}{c}\mathrm{H}\quad\mathrm{H}\\ \diagdown\ \diagup\\ \mathrm{C}\\ \diagup\ \diagdown\\ \mathrm{O}\quad\ \mathrm{O}\\ |\qquad |\\ \mathrm{M}\quad\ \mathrm{M'}\end{array} \xrightarrow{\mathrm{H2}} \begin{array}{c}\mathrm{OH}\\ |\\ \mathrm{M'}\end{array} \quad \begin{array}{c}\mathrm{OCH_3}\\ |\\ \mathrm{M}\end{array}$$

$$\mathrm{CO + MOH} \ (\uparrow) \qquad\qquad \downarrow\ \ \mathrm{CH_3OH} + \begin{array}{c}\mathrm{OH}\\ |\\ \mathrm{M'}\end{array} \quad \begin{array}{c}\mathrm{H}\\ |\\ \mathrm{M}\end{array} \qquad \text{(V)}$$

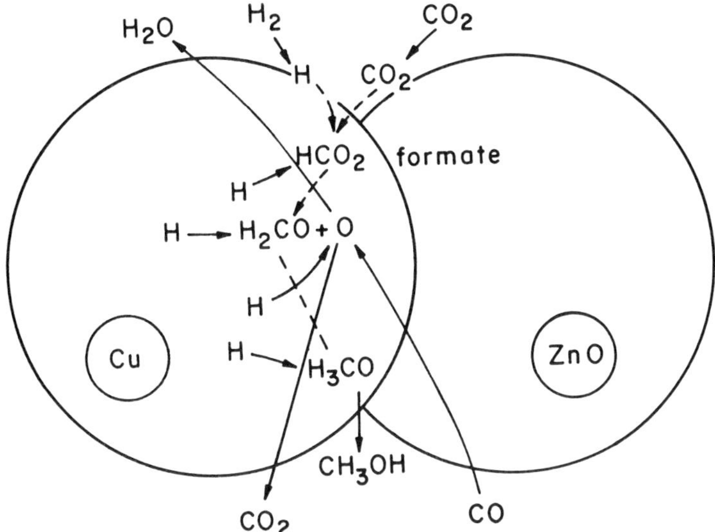

FIGURE 14. Mechanism of methanol synthesis on a $Cu/ZnO/Al_2O_3$ catalyst from a CO, CO_2, and H_2 feed. (From Chinchen, G. C., Waugh, K. C., and Whan, D. A., *Appl. Catal.*, 25, 101, 1986. With permission.)

Also, the same type of formate intermediate could be produced by the surface hydroxide and carbon monoxide. These authors concluded that the essential role of carbon dioxide may be to maintain this hydroxide intermediate, which facilitates the transformation of CO into methanol and enhances the production of methanol.

There appears to be no direct evidence to these proposals. For example, the generation of formate intermediate from CO and metal hydroxide is open to doubt. ICI researchers[110,136,143] have seriously contradicted the mechanism that involves the synthesis to proceed via CO hydrogenation. In the studies of Herman et al.,[137] the net reactant to methanol was CO (in the presence of CO_2); however, activity decayed rapidly in the absence of CO_2. Furthermore, the maximum coverage of irreversibly held CO was reported as 2% of a monolayer, despite the combined adsorption capabilities of Cu/ZnO that were significantly higher in comparison with the separate components.[113]

ICI researchers[110,136] claimed that the dominant reaction is the hydrogenation of CO_2 rather than of CO. The function of CO is likely to maintain the catalyst in the reduced state. As proposed by ICI researchers,[136] the major steps involved are Reactions 17 through 19.

$$CO_2 + 2H_2 \rightarrow CH_3OH + O(a) \tag{17}$$

$$CO + O(a) \rightarrow CO_2 \tag{18}$$

$$H_2 + O(a) \rightarrow H_2O \tag{19}$$

Their detailed mechanism of methanol synthesis is illustrated by the diagram in Figure 14.

In summary, methanol is formed through the CO_2 component of the feed, and the role of CO is to act as a reducing agent for the surface under synthesis conditions. Though the exact phase of copper in the combined catalysts under synthesis conditions is still uncertain, no unique synergy attaches to the copper/zinc oxide combination. The relative rates of Reactions 17 through 19 may control the coverage of copper surface by the intermediates (formate, formyl, and methoxy species) and oxygen.

It should be noted that the catalyst used for the methanol synthesis is also active for the water-gas shift reaction. Therefore, an alternative to Reactions 18 and 19 may be the reverse-shift reaction ($H_2O + CO \rightarrow CO_2 + H_2$) where water keeps equilibrating with CO to produce more CO_2, for the production of methanol. In addition, CO_2 in the feed gas offers a less exothermic reaction pathway (ΔH is -50 kJ/mol for the conversion of CO_2 to methanol; while, ΔH is -90 kJ/mol for the conversion of CO to methanol) for formation of methanol, and thus, helps stablize the local temperatures in the catalyst.

The specific composition of the surface intermediates on the combined catalyst has yet to be determined. However, the mechanism, as proposed by the ICI group, appears to be consistent with most of the experimental evidence available thus far, and overcomes some of the difficulties encountered by the other two mechanisms. Still, a great deal of fundamental work, perhaps utilizing high-pressure UHV technique, is required to analyze the role of Cu-ZnO catalysts in the synthesis of methanol. This will also aid in establishing the detailed kinetics of the surface processes which are virtually unknown at present. Nevertheless, it has reached the stage where the individual steps in the complex reaction network can begin to be assigned to discrete chemical entities within the catalysts. To provide complete information, however, the role of alumina should also be determined in the combined catalyst, since water is produced in the synthesis of methanol.

REFERENCES

1. **Sabatier, P. and Senderens, J. B.,** *C.R. Acad. Sci.*, 134, 514, 1902.
2. **Patart, G.,** British Filing No. 540,343, 1921.
3. **Dry, M. E.,** *Ind. Eng. Chem. Prod. Res. Dev.*, 15, 282, 1976.
4. **Fischer, F. and Tropsch, H.,** *Brennst. Chem.*, 4, 276, 1923.
5. **Fischer, F. and Tropsch, H.,** *Brennst. Chem.*, 7, 97, 1926.
6. **Dry, M. E. and Hoogendoorn, J. C.,** *Catal. Rev. Sci. Eng.*, 23, 265, 1981.
7. **Mills, G. A. and Steffgen, F. W.,** *Catal. Rev. Sci. Eng.*, 8, 159, 1973.
8. **Vannice, M. A.,** *Catal. Rev. Sci. Eng.*, 14, 153, 1976.
9. **Ponec, V.,** *Catal. Rev. Sci. Eng.*, 18, 151, 1978.
10. **Bell, A. T.,** *Catal. Rev. Sci. Eng.*, 23, 203, 1981.
11. **Anderson, R. B.,** *The Fischer-Tropsch Synthesis*, Academic Press, Orlando, Fla., 1984.
12. **Jiang, X., Stevenson, S. A., and Dumesic, J. A.,** *J. Catal.*, 91, 11, 1985.
13. **Baird, M. J. and Steffgen, F. W.,** *Ind. Eng. Chem. Prod. Res. Dev.*, 16, 142, 1977.
14. **Somorjai, G. A.,** *Catal. Rev. Sci. Eng.*, 23, 189, 1981.
15. **Dirksen, H. A. and Linden, H. R.,** *Inst. Gas. Technol. Res. Bull.*, No. 31, 1963.
16. **Seglin, L., Geosits, R., Franko, B. R., and Gruber, G.,** *Adv. Chem. Ser.*, 146, 1, 1975.
17. **Cobb, J. T. and Streeter, R. C.,** *Ind. Eng. Chem. Process Des. Dev.*, 18, 672, 1979.
18. **Blum, D. B., Sherwin, M. B., and Frank, M. E.,** *Adv. Chem. Ser.*, 146, 149, 1975.
19. **Haynes, W. P., Forney, A. J., Elliott, J. J., and Pennline, H. W.,** *Adv. Chem. Ser.*, 146, 87, 1975.
20. **Goyal, S. K.,** Ph.D. dissertation, University of Saskatchewan, Canada, 1984.
21. **Coon, V. T., Takeshita, T., Wallace, W. E., and Craig, R. S.,** *J. Phys. Chem.*, 80, 1878, 1976.
22. **Fitzharris, W. D. and Katzer, J. R.,** *Ind. Eng. Chem. Fundam.*, 17, 130, 1978.
23. **Wise, H. and Wood, B.,** U.S. Patent 4,132,672, 1979; and references cited therein.
24. **Goodman, D. W.,** *Acc. Chem. Res.*, 17, 194, 1984.
25. **Agarwal, P. K.,** Ph.D. dissertation, University of Delaware, Newark, 1979.
26. **Spencer, N. D. and Somorjai, G. A.,** *Rep. Prog. Phys.*, 46, 1, 1983.
27. **Kelley, R. D. and Goodman, D. W.,** in *The Chemical Physics of Solid Surfaces and Heterogeneous Catalysis*, King, D. A. and Woodruff, D. P., Eds., Elsevier, Amsterdam, 1982, chap. 10.
28. **Goodman, D. W.,** *J. Vac. Sci. Technol.*, 20, 522, 1982.
29. **Kolbel, H. and Roberg, H.,** *Ber. Bunsenges. Phys. Chem.*, 75, 1100, 1971.
30. **Sastri, M. V. C., Viswanathan, B., and Gupta, R.,** *J. Catal.*, 26, 212, 1972.
31. **Sastri, M. V. C., Viswanathan, B., and Gupta, R.,** *J. Catal.*, 32, 325, 1974.

32. **Kolbel, H. and Hanus, D.,** *Chem. Ing. Tech.*, 46, 1042, 1972.
33. **Kolbel, H., Ralek, M., and Jiru, P.,** *Erdoel Kohle Erdgas Petrochem. Brennst. Chem.*, 9, 580, 1970.
34. **Blyholder, G. and Neff, L. D.,** *J. Catal.*, 2, 138, 1963.
35. **Madey, T. E., Goodman, T. W., and Yates, J. T.,** 7th Int. Vacuum Congr., Vienna, 1977.
36. **Wedler, G., Papp, H., and Schroll, G.,** *J. Catal.*, 38, 153, 1975.
37. **Horgan, A. M. and King, D. A.,** in *Adsorption-Desorption Phenomena*, Ricca, F., Ed., Academic Press, London, 1972, 329.
38. **Lapujoulade, J.,** *J. Chem. Phys.*, 68, 73, 1971.
39. **Araki, M. and Ponec, V.,** *J. Catal.*, 44, 439, 1976.
40. **Wentreck, P. R., Wood, B. J., and Wise, H.,** *J. Catal.*, 43, 363, 1976.
41. **Joyner, R. W. and Roberts, M. W.,** *Chem. Phys. Lett.*, 29, 447, 1974.
42. **Joyner, R. W.,** *Far. Discuss. Chem. Soc.*, 60, 172, 1975.
43. **Kolbel, H. and Tillmetz, K. D.,** *J. Catal.*, 34, 307, 1974.
44. **Vlasenko, V. M. and Yuzefovich, G. E.,** *Russ. Chem. Rev.*, 38, 728, 1969.
45. **Farranto, R. J.,** *J. Catal.*, 41, 482, 1976.
46. **Ertl, G.,** *J. Vac. Sci.*, 14, 435, 1977.
47. **Madden, H. H. and Ertl, G.,** *Surf. Sci.*, 35, 211, 1973.
48. **Rewich, R. T. and Wise, H.,** Am. Chem. Soc. Meet., Chicago, 1977, preprint.
49. **Kummer, J. T., Dewitt, T. W., and Emmett, P. H.,** *J. Am. Chem. Soc.*, 70, 3632, 1948.
50. **King, D. L.,** *J. Catal.*, 61, 77, 1980.
51. **Eckerdt, J. G. and Bell, A. T.,** *J. Catal.*, 58, 170, 1979.
52. **Tamaru, K.,** Preprints, 7th Int. Congr. Catal., Tokyo, 1980.
53. **Falconer, J. L. and Zagli, E.,** *J. Catal.*, 62, 280, 1980.
54. **Zagli, E. and Falconer, J. L.,** *J. Catal.*, 69, 1, 1981.
55. **Trimm, D. L.,** *Trans. Inst. Chem. Eng.*, 54, 119, 1976.
56. **McCarty, J. G. and Wise, H.,** *J. Catal.*, 57, 406, 1979.
58. **Bileon, P., Helle, J. N., and Sachtler, W. M. H.,** *J. Catal.*, 58, 95, 1979.
59. **McCarty, J. G., Hou, P. Y., Sheridan, D., and Wise, H.,** 7th North Am. Meet. Catal. Soc., Boston, 1981.
60. **Bartholomew, C. H.,** *Catal. Rev. Sci. Eng.*, 24, 67, 1982.
61. **Kieffer, E. P. and Van der Baan, H. S.,** *Appl. Catal.*, 3, 245, 1985.
62. **Ino, T., Watanbe, H., and Morita, Y.,** *J. Jpn. Petrol. Inst.*, 24, 181, 1981.
63. **Bartholomew, C. H. and Vance, C. K.,** *J. Catal.*, 91, 78, 1985.
64. **Kuijpers, E. G. M. and Geus, J. W.,** *Fuel*, 62, 158, 1983; Kuijpers, E. G. M., Jansen, J. W., Van Dillen, A. J., and Geus, J. W., *J. Catal.*, 72, 75, 1981; Kuijpers, E. G. M., Breedjik, A. K., Van der Wal, W. J. J., and Geus, J. W., *J. Catal.*, 72, 210, 1981.
65. **Castner, D. G., Blackader, R., and Somorjai, G. A.,** *J. Catal.*, 66, 257, 1980.
66. **Vannice, M. A.,** *J. Catal.*, 44, 152, 1976.
67. **Dalla Betta, R. A., Piken, A. G., and Shelef, M.,** *J. Catal.*, 40, 173, 1975.
68. **Fontaine, R.,** Ph.D. dissertation, Cornell University, Ithaca, N. Y., 1973.
69. **Schoubye, P.,** *J. Catal.*, 14, 238, 1969.
70. **Bousquet, J. L. and Teichner, S. J.,** *Bull. Soc. Chim. Fr.*, 2963, 1969.
71. **Agarwal, P. K., Katzer, J. R., and Manogue, W. H.,** *J. Catal.*, 69, 312, 1981.
72. **Vannice, M. A.,** *J. Catal.*, 37, 449, 1975.
73. **Karn, F. S., Shultz, J. F., and Anderson, R. G.,** *Ind. Eng. Chem. Prod. Res. Dev.*, 4, 265, 1965.
74. **Quach, T. Q. P. and Rouleau, D.,** *J. Appl. Chem. Biotechnol.*, 28, 378, 1978.
75. **Ozdogan, S. Z., Gochis, P. D., and Falconer, F. D.,** *J. Catal.*, 83, 257, 1983.
76. **Rabo, J. A., Elek, L. F., and Francis, J. N.,** *Stud. Surf. Sci. Catal.*, 7, 490, 1981.
77. **Goodman, D. W., Yates, J. T., and Madey, T. E.,** *Surf. Sci.*, 93, L135, 1980.
78. **Goodman, D. W., Yates, J. T., and Madey, T. E.,** *J. Catal.*, 50, 279, 1979.
79. **Yates, J. T. and Madey, T. E.,** *J. Chem. Phys.*, 54, 4969, 1971.
80. **Vorburger, T. V., Sandstorm, D. R., and Waclawski,** *Surf. Sci.*, 6, 211, 1976.
81. **Benziger, J. and Madix, R. J.,** *Surf. Sci.*, 77, L379, 1978.
82. **Froitzheim, H., Ibach, H., and Lehwald, S.,** *Surf. Sci.*, 63, 56, 1977.
83. **Sexton, B. A. and Somorjai, G. A.,** *J. Catal.*, 46, 167, 1977.
84. **Krebs, H. J., Bonzel, H. P., and Gafner, G.,** *Surf. Sci.*, 88, 269, 1979.
85. **Dwyer, D. J. and Somorjai, G. A.,** *J. Catal.*, 52, 291, 1978.
86. **Goodman, D. W., Kelley, R. D., Madey, T. E., and Yates, J. T.,** *J. Catal.*, 63, 226, 1980.
87. **Kelley, R. D. and Goodman, D. W.,** Am. Chem. Soc. Div. Fuel Chem. Meet., Houston, 1980, preprint.

88. **Goodman, D. W. and White, J. M.,** *Surf. Sci.,* 90, 201, 1979.
89. **Kelley, R. D., Madey, T. E., Revesz, A., and Yates, J. T.,** *Appl. Surf. Sci.,* 1, 266, 1978.
90. **Yates, J. T., Gates, S. M., and Russell, J. N.,** *Surf. Sci.,* 164, L839, 1985.
91. **Bishke, S. D., Goodman, D. W., and Falconer, J. L.,** *Surf. Sci.,* 150, 351, 1985.
92. **Richardson, J. T.,** *J. Catal.,* 21, 130, 1971.
93. **Rostrup-Nielsen, J. R.,** *J. Catal.,* 21, 171, 1971.
94. **Oudar, J.,** *Catal. Rev. Sci. Eng.,* 22, 171, 1980.
95. **Fitzharris, W. D., Katzer, J. R., and Monogue, W. H.,** *J. Catal.,* 76, 369, 1982.
96. **Rostrup-Nielsen, J. R. and Pederson, K.,** *J. Catal.,* 59, 395, 1979.
97. **Ng, C. F. and Martin, G. A.,** *J. Catal.,* 54, 385, 1978.
98. **Rewick, R. T. and Wise, H.,** *J. Phys. Chem.,* 82, 157, 1978.
99. **Ponnell, R. B., Chung, K. S., and Bartholomew, C. H.,** *J. Catal.,* 46, 340, 1977.
100. **Hagstrum, H. D. and Becker, G. E.,** *J. Chem. Phys.,* 54, 1015, 1971.
101. **Fisher, G. B.,** *Surf. Sci.,* 62, 31, 1977.
102. **Goodman, D. W. and Kiskanova, M.,** *Surf. Sci.,* 105, L265, 1981.
103. **Kiskanova, M. and Goodman, D. W.,** *Surf. Sci.,* 108, 64, 1981.
104. **Houston, J. W., Rogers, J., Goodman, D. W., and Belton, D. N.,** *J. Vac. Sci. Technol.,* A2, 882, 1984.
105. **Goodman, D. W.,** *Appl. Surf. Sci.,* 19, 1, 1984.
106. **Kiskanova, M. and Goodman, D. W.,** *Surf. Sci.,* 109, L555, 1981.
107. **Peebles, D. W., Goodman, D. W., and White, J. M.,** *J. Phys. Chem.,* 87, 4378, 1983.
108. **Bartholomew, C. H., Agarwal, P. K., and Katzer, J. R.,** *Advances in Catalysis,* Vol 31, Eley, D. D., Pines, H., and Weisz, P. B., Ed., Academic Press New York, 1982, 136.
109. **Stiles, A. B.,** *AIChE J.,* 23, 362, 1977.
110. **Bowker, M.,** *Vacuum,* 33, 669, 1983.
111. **Marschner, F. and Moeller, F. W.,** in *Applied Industrial Catalysis,* Vol 2, Leach, B. E., Ed., Academic Press, New York, 1983, chap. 6.
112. **Kung, H. H.,** *Catal. Rev. Sci. Eng.,* 22, 235, 1980.
113. **Klier, P.,** *Advances in Catalysis,* Vol 31, Eley, D. D., Pines, H., and Weisz, P., Eds., Academic Press, New York, 1982, 243.
114. **Andrew, S. P. S.,** Post Congr. Symp. Int. Congr. Catal. 7th, Osaka, paper 12, 1980.
115. **Sugier, A. and Freund, E.,** U.S. Patent 4,122,110, 1978.
116. **Poutsma, M. L., Elek, L. F., Ibarbia, D. A., Risch, A. P., and Rabo, J. A.,** *J. Catal.,* 52, 157, 1978.
117. **Ichikawa, M.,** *Bull. Chem. Soc. Jpn.,* 51, 2268, 1978.
118. **Hicks, R. F. and Bell, A. T.,** *J. Catal.,* 91, 104, 1985.
119. **Ryndin, Y. A., Hicks, R. F., Bell, A. T., and Yermakov, Y. I.,** *J. Catal.,* 70, 287, 1981.
120. **Poels, E. K., van Broekhoven, E. H., van Barneveld, W. A. A., and Ponec, V.,** *React. Kinet. Catal. Lett.,* 18, 223, 1981.
121. **Fajula, F., Anthony, R. G., and Lunsford, J. H.,** *J. Catal.,* 73, 237, 1982.
122. **Kikuzono, Y., Kagami, S., Naito, S., Onishi, T., and Tamaru, K.,** *Faraday Discuss. Chem. Soc.,* 72, 135, 1982.
123. **Ponec, V.,** *Stud. Surf. Sci. Catal.,* 11, 63, 1982.
124. **Driessen, J. M., Poels, E. K., Hinderman, J. P., and Ponec, V.,** *J. Catal.,* 82, 26, 1983.
125. **Bradley, J. S.,** *J. Am. Chem. Soc.,* 107, 7419, 1979.
126. **Rathke, J. W. and Feder, H. M.,** *J. Am. Chem. Soc.,* 100, 3627, 1978.
127. **Odebunmi, E. O., Zhao, Y., Knozinger, H., Tesche, B., Manogue, W. H., Gates, B. C., and Hulse, J.,** *J. Catal.,* 86, 95, 1984.
128. BNL Catalyst, Chem. Eng. News, p. 21, Aug. 1986.
129. **Bardet, R. and Trambouze, Y.,** *J. Chim. Phys. (Paris),* 78, 135, 1981.
130. **Rozovskii, A. Y.,** *Khim. Prom. (Kiev),* 11, 652, 1980.
131. **Rozovskii, A. Y., Lim, G., Liberov, L. B., Sliinkii, E. V., Loktev, S. M., Kagan, Y. B., and Baskhirov,** *Kinet. Catal. (USSR),* 18, 691, 1977.
132. **Popov, I. G., Lender, Y. V., Karavaev, M. M., and Vislogusova, V. G.,** *Khim. Prom. (Kiev),* 6, 420, 1974.
133. **Denny, P. J. and Whan, D. A.,** *Catalysis (London),* 2, 46, 1978.
134. **Liu, G., Wilcox, D., Garland, M., and Kung, H.,** *J. Catal.,* 90, 139, 1984.
135. **Chinchen, G. C., Denny, P. J., Parker, D. G., Short, G. D., Spencer, M. S., Waugh, K. C., and Whan, D. A.,** *Am. Chem. Soc. Div. Fuel Chem.,* 29(5), 178, 1984.
136. **Chinchen, G. C., Waugh, K. C., and Whan, D. A.,** *Appl. Catal.,* 25, 101, 1986.
137. **Herman, R. G., Klier, K., Simmons, G. W., Finn, B. P., Bulko, J. B., and Kobylinski, T. P.,** *J. Catal.,* 56, 407, 1979.

138. **Mehta, S., Simmons, G. W., Klier, K., and Herman, R. G.,** *J. Catal.,* 57, 339, 1979.
139. **Bulko, J. B., Herman, R. G., Klier, K., and Simmons, G. W.,** *J. Phys. Chem.,* 83, 3118, 1979.
140. **Friedrich, J. B., Wainwright, M. S., and Young, D. J.,** *J. Catal.,* 80, 1, 1983.
141. **Marsden, W. L., Wainwright, M. S., and Friedrich, J. B.,** *Ind. Eng. Chem. Prod. Res. Dev.,* 19, 551, 1980.
142. **Fleisch, T. H. and Mieville, R. L.,** *J. Catal.,* 90, 165, 1984.
143. **Chinchen, G. C. and Waugh, K. C.,** *J. Catal.,* 97, 280, 1986.
144. **Natta, G. N., Pino, P., Mizzanti, G., and Pasquon, I.,** *Chim. Ind. Milan,* 35, 705, 1953.
145. **Uchida, H. and Ogino, Y.,** *Bull. Chem. Soc. Jpn,* 31, 45, 1958.
146. **Leonov, V. E., Karavaev, M. M., Tsybina, E. N., and Petrischcheva, G. S.,** *Kinet. Katal.,* 14, 848, 1973.
147. **Bakermeir, H., Laurer, P. R., and Schroder, W.,** *CEP Symp. Ser.,* 66, 1, 1970.
148. **Tsuchiya, S. and Shiba, T.,** *Bull. Chem. Soc. Jpn,* 38, 1726, 1965.
149. **Klier, K., Chatikavanij, V., Herman, R. G., and Simmons, G. W.,** *J. Catal.,* 74, 343, 1982.
150. **Hayward, D. and Trapnell, B.,** *Chemisorption,* Butterworths, London, 1964, 164.
151. **Horn, K., Hussain, M., and Pritchard,** *J. Surf. Sci.,* 63, 244, 1977.
152. **Balooch, M., Cardillo, M., Miller, D., and Stickney, M.,** *Surf. Sci.,* 46, 358, 1974.
153. **Bowker, M. and Madix, R. J.,** *Surf. Sc.,* 95, 190, 1980.
154. **Wachs, I. E., and Madix, R. J.,** *J. Catal.,* 53, 208, 1978.
155. **Marien, J., De Pauw, E., and Peizer, G.,** *J. Phys. Chem.,* 87, 4344, 1983.
156. **Natta, G.,** *Catalysis,* Vol. 3, Emmet, P. H., Ed., Reinhold, New York, 1955.
157. **Bowker, M., Houghton, H., Waugh, K. C., Giddings, T., and Green, M.,** *J. Catal.,* 84, 252, 1983.
158. **Gopel, W.,** *Surf. Sci.,* 62, 165, 1977.
159. **Chang, S. C. and Mark, P.,** *Surf, Sci.,* 45, 721, 1974.
160. **Lanks, I. and Green, M.,** *Surf. Sci.,* 71, 735, 1975.
161. **Grunze, M., Hischwald, W., and Hofmann, D.,** *J. Crystal Growth,* 52, 241, 1981.
162. **Ueno, A., Onishi, T., and Tamaru, K.,** *Trans. Faraday Soc.,* 66, 756, 1970; 67, 3585, 1971.
163. **Bowker, M., Houghton, H., and Waugh, K. C.,** *J. Chem. Soc. Faraday Trans. 1,* 77, 3023, 1981.
164. **Herman, R. G., Simmons, G. W., and Klier, K.,** *Proc. 7th Int. Congr. Catal.,* Seiyama, T. and Tanabe, K., Eds., Kodansha, Tokyo, 1981, 475.
165. **Okamoto, Y., Fukino, K., Imanaka, T., and Teranishi, S.,** *J. Phys. Chem.,* (a) 87, 3740, 1983; (b) 87, 3747, 1983.
166. **Apai, G. R., Monnier, J. R., and Hanrahan, M J.,** *J. Chem. Soc. Chem. Commun.,* 212, 1984.
167. **Denise, B. and Sneeden, R. P. A.,** *Chem. Technol.,* 12, 108, 1982.
168. **Denise, B., Sneeden, R. P. A., and Hamon, C.,** *J. Mol. Catal.,* 17, 359, 1982.
169. **Biloen, P. and Sachtler, W. M. H.,** *Adv. Catal.,* 30, 165, 1981.
170. **Ponec, V.,** in *Catalysis, Specialist Periodical Reports,* Bond, G. C. and Webb, G., Eds., Royal Society of Chemistry, London, 1982, 48.
171. **Barteau, M. A., Fuelner, P., Stengl, R., Broughton, J. Q., and Menzel, D.,** *J. Catal.,* 94, 51, 1985.
172. **Lauderback, L. L. and Delgass, W. N.,** Symp. on Role of Solid State Chemistry in Catalysis, Div. of Petr. Chem., Am. Chem. Soc., Washington, D.C., 1983, preprint.
173. **Kaminsky, M. P., Winograd, N., Geoffroy, G. L., and Vannice, M. A.,** *J. Am. Chem. Soc.,* 108, 1315, 1986.
174. **Vohs, J. M. and Barteau, M. A.,** *Surf. Sci.,* in press.
175. **Zweicker, G., Jacobi, K., and Cunningham, A.,** *Int. J. Mass Spectrom. Ion Phys.,* 60, 213, 1984.
176. **Akhter, S., Lui, K., and Kung, H. H.,** *J. Phys. Chem.,* 89, 1958, 1984.
177. **Madix, R., Thornburg, M., and Lee, S. B.,** *Surf. Sci.,* 13, L447, 1983.

Chapter 6

REFORMING

I. GENERAL

One of the major landmarks in the history of petroleum refining is the catalytic reforming process — the processing of virgin and cracked naphthas using several reactor beds of catalysts at high temperature and moderate pressure to produce high-octane gasolines. The naphtha (C_5 — 477 K) can be obtained directly from the crude unit or from the fractioned product of another refinery process, such as a coking. Catalysts for reforming reactions are small crystallites of Pt or Pt clusters, generally a few tenths percent platinum (in admixture with other noble metals and a halogen) supported on porous alumina; the catalysts are considered to be bifunctional, since both the metal and oxide play active roles.

During the last 3 decades, several processes evolved very rapidly. The basic differences between these processes are the catalyst and the catalyst regeneration process. While some processes are nonregenerative, others use intermittent or cyclic regeneration or a combination thereof. Another distinguishing feature is whether they use a fixed bed or a fluidized bed for reforming reactions. The latest development is a design in which the catalyst moves continuously through the reactor, and finally to a regeneration vessel.[1]

In addition to changes in the process design, the catalyst used in reforming has been modified to offer improved performance in terms of higher reformate yield and especially stability with respect to deactivation. The most significant development, which occurred in the late 1960s, is the bimetallic (platinum-rhenium) reforming catalyst, which displays a slower rate of deactivation than the one-component catalyst. In fact, catalytic reforming was among the first industrial processes for which multifunctional catalysts were knowingly created.[2]

Since the early 1970s, platinum has always been combined with other metals such as rhenium, iridium, and tin when used as a reforming catalyst.[3] However, the technological advantages of these bimetallic catalysts are counterbalanced by the complexity of their structures. The function of Re is not well understood, and there is still controversy about the presence of Re in a metallic state (a cluster with Pt) or in an ionic state on the surface.[4,5] The full characterization of bimetallics is a challenge that will demand the innovative efforts of scientists and engineers for decades to come.

This chapter proceeds with an overview of the reforming process (feedstock, reactions, and process description) and follows with detailed accounts of the reforming catalysts, reforming reactions, and reaction engineering. Since research on bimetallic catalysts has had a major impact in the reforming of petroleum naphtha fractions, this subject is, therefore, developed in detail. The bimetallic catalysts of the type practiced commercially will be considered, and some potential applications not currently in use will be introduced.

II. PROCESS

A. Reactions

Naphtha reforming is a complex process involving reactions of a large number of hydrocarbons. Among the overwhelming number of reactions that occur during catalytic reforming, the main reactions are naphthene dehydrogenation, naphthene isomerization, dehydrocyclization, paraffin isomerization, and hydrocracking. Examples of each of these reactions are given in Figure 1. These reactions occur to varying degrees, and the extent to which each

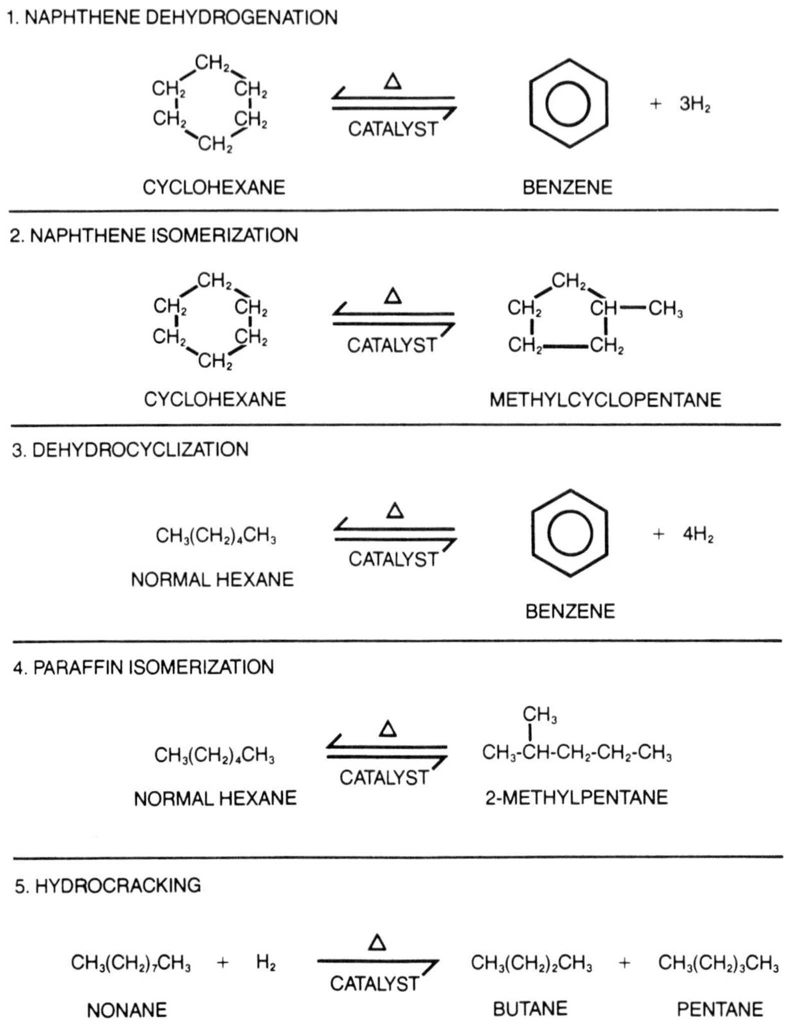

1. NAPHTHENE DEHYDROGENATION

CYCLOHEXANE ⇌ (Δ, CATALYST) BENZENE + $3H_2$

2. NAPHTHENE ISOMERIZATION

CYCLOHEXANE ⇌ (Δ, CATALYST) METHYLCYCLOPENTANE

3. DEHYDROCYCLIZATION

$CH_3(CH_2)_4CH_3$ ⇌ (Δ, CATALYST) BENZENE + $4H_2$

NORMAL HEXANE

4. PARAFFIN ISOMERIZATION

$CH_3(CH_2)_4CH_3$ ⇌ (Δ, CATALYST) $CH_3\text{-}CH(CH_3)\text{-}CH_2\text{-}CH_2\text{-}CH_3$

NORMAL HEXANE — 2-METHYLPENTANE

5. HYDROCRACKING

$CH_3(CH_2)_7CH_3 + H_2$ → (Δ, CATALYST) $CH_3(CH_2)_2CH_3 + CH_3(CH_2)_3CH_3$

NONANE — BUTANE — PENTANE

FIGURE 1. Examples of reforming reactions. (From Edgar, M. D., in *Applied Industrial Catalysis,* Vol 1, Leach, B. E., Ed., Academic Press, New York, 1983, chap. 5. With permission.)

takes place depends upon the nature of the catalyst, the composition of the naphtha feed, and the conditions of operation. Table 1 gives the octane numbers of some of the pure hydrocarbons which indicate the types of hydrocarbons desired for high-octane gasoline.[6]

It has been shown that the isomerization and hydrocracking reactions require the catalyst to have two separate and distinct functions — a hydrogenation-dehydrogenation function and an acidity function. These may reside either in different chemical components or in the same component. This is one of the distinctions which separates the metal-acidic oxide catalysts from the metal oxides. In the former, an active metal, Pt, serves as the catalytic site for hydrogenation-dehydrogenation and metal-catalyzed hydrocracking reactions. The acidic component (halogenated alumina) provides the acid sites for the isomerization, cracking, and polymerization reactions of olefins. This is achieved by treating the alumina support with HCl, whereby some of the surface hydroxyl groups react to give chloride ions on the surface. However, the chlorided alumina by itself is not sufficiently active for paraffin isomerization. The key to the success lies in the interaction between the metal and acid

Table 1
OCTANE NUMBER OF PURE HYDROCARBONS

Hydrocarbon	Research octane number (clear)
Paraffins	
n-Butane	113
n-Pentane	62
n-Hexane	19
n-Heptane	0
n-Octane	−19
2-Methylhexane	41
Naphthenes	
Methylcyclopentane	107
Cyclohexane	110
Methylcyclohexane	104
Aromatics	
Benzene	99
Toluene	124

functions.[7,8] The Pt catalyzes dehydrogenation of paraffins to form olefins which become key intermediates since they easily protonate to carbenium ions on the acidic sites, and thereby are isomerized. Exothermic temperature effects associated with both naphthene and paraffin isomerizations are usually small and go undetected in a refinery reformer. On the other hand, hydrocracking is exothermic and normally occurs in the last reactor.

An interesting and very useful property of platinum is that, with the acid portion of the catalyst, it can be made to catalyze selectively the slower turnover reactions, such as dehydrocyclization, an important octane-enhancing reaction in which paraffins are converted to aromatics; the hydrogenation-dehydrogenation pathway may be virtually switched off. This endothermic reaction usually occurs in the middle of the last reactors of the reformer unit, and its rate is inhibited by hydrogen partial pressure.

Naphthene dehydrogenation is a highly endothermic and fast reaction in which naphthenes are converted to aromatics. Typical feeds contain 18 to 50% naphthenes as cyclopentanes and cyclohexanes. Most of the naphthene dehydrogenation is completed in the first reactor of the reformer. The thermodynamics of the dehydrogenation reactions require operation at low pressures and high temperatures. These conditions are quite favorable for rapid coke formation. Coke is unavoidably deposited rapidly on the entire catalyst, especially when the hydrogen partial pressure is low. Platinum has generally been replaced by Pt-Re and/or Pt-Ir because of its resistance to deactivation by coke formation.

B. Feedstock

Straight-run naphthas generally constitute the major feedstocks to reformers. Characteristics of three examples of naphtha feeds are given in Table 2. The composition of the stock has a major effect on the reactor temperature required to achieve a desired product octane value and on the reformate yield. The higher the paraffin content of the feed, the more difficult it is to reform. For example, light Arabian naphtha requires higher reactor temperatures, and yields less reformate in comparison to midcontinent naphtha. Also, stocks which have an appreciable concentration of unsaturated hydrocarbons must usually be hydrotreated before reforming in order to prevent undue hydrogen consumption in the reformer and excessive catalyst deactivation. Among the various ways to rank naphtha as to reformability, one method uses a value calculated by adding the naphthene content to twice the aromatic content and is referred to N + 2A value.[1] In Table 2 are N + 2A values of the three naphthas. Until the late 1970s, N + 2A values of 60 to 65 were typical of naphthas

Table 2
FEEDSTOCK EXAMPLES

Naphtha properties	Light Arabian naphtha	Midcontinent naphtha	Persian Gulf naphtha
Density (g/mℓ)	0.715	0.755	0.741
ASTM distillation			
Boiling range (K)	353—423	353—439	343—463
Composition (vol%)			
Paraffins	74	45	70
Naphthenes	19	45	18
Aromatics	7	10	12
N + 2A	33	100	42

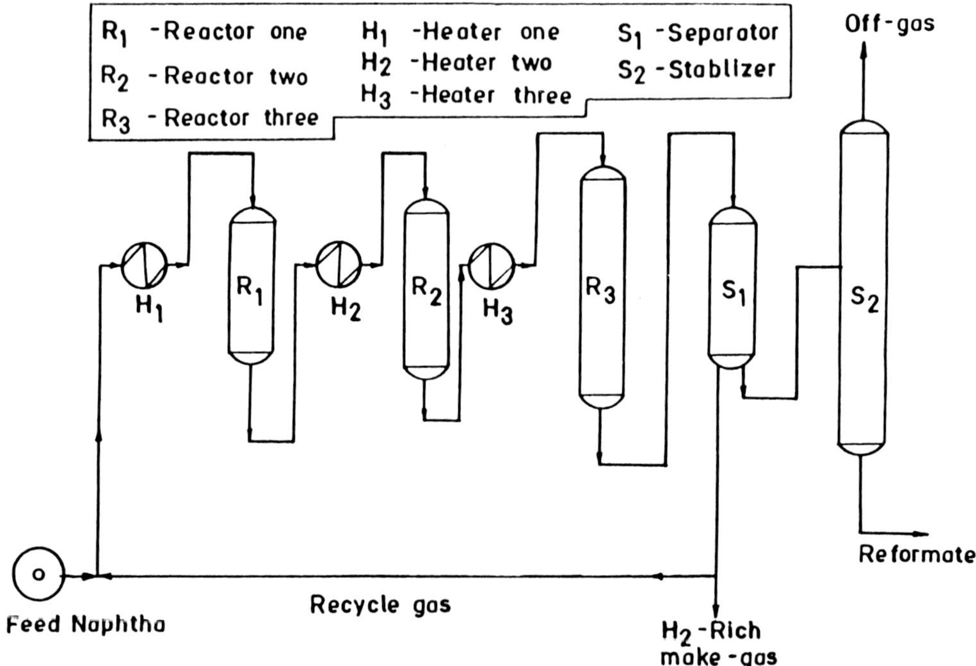

FIGURE 2. Process flow diagram for three reactor reforming unit.

being run by the U.S. refiners. In the 1980s, refiners were required to use less desirable crudes, and the rank of naphthas began to decrease. Therefore, naphthas similar to the light Arabian naphtha in N + 2A value are becoming more common.

C. Process Description

Most commercial reformers consist of multi-bed reactors, pre- and reheaters, separators, and stabilizers. A block diagram for a three-reactor reforming unit is shown in Figure 2. Usual feedstocks are napththas in the boiling range 350 to 500 K. The reformers are operated at relatively high pressures (25 to 50 atm); temperatures in the range 740 to 780 K and high hydrogen to hydrocarbon mole ratios of 5:10 are employed. The recent development of much stabler reforming catalysts has increased interest in low-pressure reforming.

Since 1949, the typical monometallic catalyst has been platinum supported on chlorided alumina, $Pt/A1_2O_3$. The metallic function of the catalyst was later improved by the addition of a second precious metal; and, in 1968, Kluksdahl[9] obtained the patent (assigned to

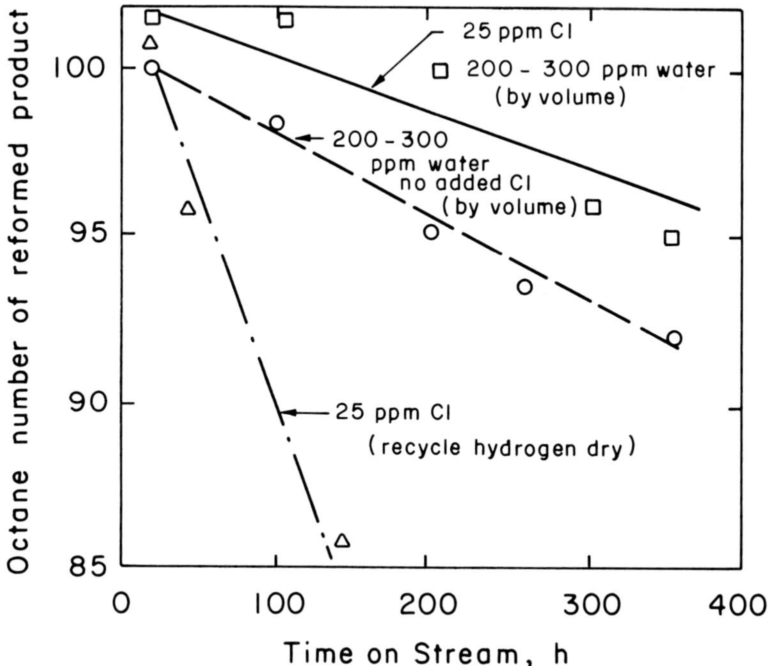

FIGURE 3. Effect of feed chlorine content and recycle hydrogen stream water content on reforming catalyst activity and stability. Conditions of pilot plant study: Temperature 769 K, LHSV = 3.0 hr^{-1}, H_2/HC = 6, Pressure = 150 atm. (From Svajgl, O., *Int. Chem. Eng.*, 12(1), 55, 1972. Reproduced by permission of the American Institute of Chemical Engineers.)

Chevron) for the addition of rhenium; since then, platinum is always clustered with a second metal, such as Ir,[10] Ge,[11] or Sn.[12,13] The two most commonly used bimetallic reforming catalysts in commercial reformers are Pt-Re on alumina and Pt-Ir on alumina. The latter has received considerable importance, most notably from researchers at Exxon.[10] Tin also continues to receive attention as a modifier metal to stabilize activity of Pt naphtha-reforming catalysts. Some plants still use a monometallic catalyst, where the supply of naphtha feedstocks is of high quality, the severity of the reforming process can be kept low. In such cases, the deactivation of the catalyst is slow and so is the temperature increase needed to counteract the loss in product quality.

In commercial catalysts, platinum concentration ranges between 0.1 and 1.0% with 0.3 to 0.7% being the preferred range. Halogen, 0.1 to 1.0%, (generally chlorine) is added to the alumina base to provide the necessary acidity. The alumina for support is one of two crystalline forms — eta or gamma. Although the eta form has a higher acid function than the gamma form, the γ-alumina is thermally stable and retains more of its initial surface through repeated use and regeneration.[1,14] The lower acid function of the γ-alumina catalyst can be compensated for by proper adjustment of the halogen content of the catalyst. Also, since Cl^- is continually stripped from the surface as HCl by reaction with small amounts of water in the feed (or water produced from oxygen in the feed), the Cl content of the catalyst must be maintained by adding chlorinated organic compounds to the feed; the concentrations are specified to fix the ratio of Cl to H_2O.[15,16] Figure 3 illustrates how the effects are balanced with the proper ratio of Cl to water in the feed. The balance of water-chloride is usually specific for each catalyst type and therefore considered proprietary information by the process licensor.

To decrease the gas formation (methanation reactions and excessive hydrocracking), sul-

Table 3
TYPICAL OPERATING CONDITIONS FOR A THREE-REACTOR SYSTEM

	Reactor		
	1	2	3
Inlet temp., K	775	775	775
Exit temp., K	706	744	769
Temp. drop, K	342	304	279
Octane number (clear)	65	80	90
Octane number increase	27	14	10
LHSV, h^{-1} per reactor	5.5	2.4	1.7
% total catalyst charge	15	35	50
Principal reactions			

Reactor 1: Dehydrogenation and dehydroisomerization
Reactor 2: Dehydrogenation, dehydroisomerization, hydrocracking, and dehydrocyclization
Reactor 3: Hydrocracking and dehydrocyclization

From Gates, B. C., Katzer, J. R., and Schuit, G. C. A., *Chemistry in Catalytic Processes,* McGraw-Hill, New York, 1979. With permission.

furization of the catalyst is necessary (after reduction of the catalyst but before the introduction of feedstocks). It also prevents a thermal runaway associated with a high rate of cracking. Following reduction, a typical procedure might call for 0.06 wt% sulfur to be introduced into the individual reactors.[1] Hydrogen sulfide has been widely employed for this purpose, though dimethyl sulfide can also be used due to easier handling.

The overall reforming process is endothermic, which suggests that the temperature of the fixed-bed reactor decreases in the direction of flow. As a result, the size of the first reactor is small compared to the size of the end reactor. Also, because of the rapid cooling of the reactant stream resulting from the endothermic dehydrogenation reactions, the residence time and the amount of catalyst in the first (upstream) reactor is considerably less than each of the other reactors. A typical distribution of operating conditions (temperature, liquid hourly space velocity [LHSV], octane number distribution based on octane number of feed as 38.5, catalyst charge, etc.) for a three-reactor system is shown in Table 3.[8]

A major feature that merits attention in a catalytic reforming process is the eventual decline of catalytic activity due to carbon deposition and the consequent need for regeneration which forms the distinguishing feature of many commercial processes. Regeneration, in general, involves burning the coke off the catalyst under carefully controlled conditions in order to restore a reforming catalyst to its original activity. The three major categories of regenerative types are[17]

1. Periodic regeneration
2. Semicontinuous regeneration
3. Continuous regeneration

Each is associated with a particular range of operating conditions.

To counteract the product quality decline resulting from the loss of catalytic activity, the operating temperatures are gradually increased. When it is no longer practical to increase the temperature, the whole plant is shut down and the catalyst is regenerated. It is apparent that as the severity of the operating conditions increases, the rate of activity decline also increases leading to frequent shut downs.

In the semicontinuous regenerative process, an additional 'swing' reactor is used which replaces one of the others when its catalyst charge needs to be regenerated. As a result, there are no major shut downs. But the drawbacks of this process are the need for a complex layout and switching policy and the requirement that all reactors be of the same size to make switching among them possible.

In a moving-bed process, small quantities of the catalyst are continuously withdrawn from an operating reactor, transported to a regeneration unit, regenerated, and returned to the reactor system. It consists of stacked reactors allowing catalyst transport through the reactors via gravity flow. The catalyst is withdrawn from the bottom reactor, comparable to the third or fourth reactor in a conventional design, and the regenerated catalyst is returned to the top reactor, equivalent to the first reactor.

Although cycle lengths are highly variable and dependent upon most of the operating conditions of the unit, catalysts, and feedstock quality, most units would run from 6 months to 1 year between regenerations, with a total life of at least 10 cycles before being replaced. Bimetallic catalysts have lasted for over 10 years of operation.[1]

Regeneration procedure typically is an oxidation process. Some refiners introduce oxygen to both the first and last reactors at the same time with a gas containing from 0.5 to 1% O_2. In a semiregenerative unit, regeneration at low pressure (approximately 8 atm) with air as the source of oxygen is the most common method. However, in any case, the maximum bed temperature should not exceed 725 K, in order to prevent damage to the catalyst, particularly sintering and loss of surface area of the Pt component.

Finally, disturbances in the feed pretreatment or contaminants in the feed can poison the reforming catalyst during processing; sulfur, nitrogen, and chloride are temporary poisons and lead and arsenic are permanent ones. The catalyst can only tolerate few parts per billion. For example, nitrogen attacks the acidic sites by forming an end product as ammonium chloride.

In the following pages, an overview of the reforming with Pt-Al_2O_3 catalyst is presented as this forms the main component of the bimetallic catalysts. Thus, this portion of the text is expected to form a good background for studies related to bimetallic catalysts.

III. NATURE OF Pt-Al_2O_3 CATALYSTS IN REFORMING REACTIONS: AN OVERVIEW

The platinum-alumina catalyst system belongs to the class of catalysts known as bifunctional catalysts.[18] These catalysts consist characteristically of two principal components: a metal (Pt) dispersed on an acidic support (Al_2O_3). Pt is well known to be highly active as a catalyst for hydrogenation and dehydrogenation reactions. The role of the support is to accelerate acid-catalyzed reactions such as isomerization and cyclization. Because the support alone possesses activity and may interact synergistically with the metal component, the activity and selectivity of bifunctional catalysts is often much different from that of catalysts possessing purely metallic properties.

The bifunctional behavior of Pt-Al_2O_3 catalysts during catalytic reforming has been discussed in detail by Ciapetta and Wallace in their 1971 review.[18] Reactions of hydrocarbons on Pt have also been studied extensively. Much of the discussion has taken place in the context of hydrocarbon conversion reactions related to naphtha reforming. An extensive journal literature dealing with model reactions is available, as well as several patents describing full-scale reforming studies over an extended period. Of the reviews and books published, those by the following are noteworthy: Sinfelt,[3] Gates et al.,[8] Anderson,[19] Clarke and Rooney,[20] Gault,[21] and Ponec.[22] It is not our intention to repeat all these results, including those already reviwed, but to focus on those points which help us to rationalize the data obtained with bimetallics. The section begins by recalling a few features of reforming catalytic

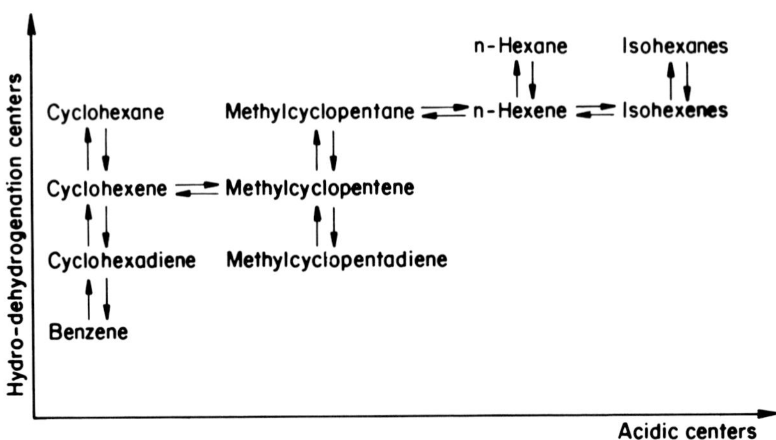

FIGURE 4. Reaction network for reforming of C_6 hydrocarbons. (Reprinted with permission from Mills, G. A., Heinmann, H., Milliken, T. H., and Obland, A. G., *Ind. Eng. Chem.*, 45, 134, Copyright 1953, American Chemical Society.)

chemistry; a review[23] appeared about 25 years ago and interestingly, its conclusions still prevail.

A. Mechanistic Considerations

The two functions of the reforming reactions interact through olefins, which are the key intermediates in the reaction network. This was originally proposed by Mills and co-workers.[24] The scheme is shown in Figure 4. The vertical reaction paths in the figure take place on the hydrogenation-dehydrogenation centers of the catalyst and the horizontal reaction paths on the acidic centers. According to the mechanism, the conversion of *n*-hexane to benzene, for example, first involves dehydrogenation on the metal to give straight-chain hexene. The hexene migrates to a neighboring acid center; there it is protonated to give a secondary carbonium ion, which can react to form methylcyclopentane, which can react further to form cyclohexene and then benzene. Alternatively, the secondary carbonium ion can isomerize and desorb as isohexene and migrate to the metal function, where it can be adsorbed and hydrogenated to given isohexane. The relative amounts of these products depend upon the reaction conditions.

A few years later, Weisz and co-workers[23] demonstrated that the metal and acidic oxide can, in fact, act independently, in that mechanical mixtures of the two components contained in separate particles were able to catalyze the isomerization of *n*-paraffins to branched paraffins at a rate comparable to that obtained in conventional dual-function catalysts, while the individual components were not active for this reaction. Hindin et al.[25] also demonstrated that mechanical mixtures of this type were active in the dehydroisomerization of methylcyclopentane to benzene. Thus, it is evident that the two elements of dual-function catalysts retain their identities, and reactions requiring their combined action proceed through reaction intermediates which can be transported through gas phase.[26]

Several researchers[27-29] found that these platinum catalysts catalyzed almost all reactions which are known to occur with Pt-Al_2O_3 under reforming. While it has been shown that a purely metal-catalyzed isomerization process can occur, this path is probably not an important challenge to the commonly accepted mode of action of bifunctional catalysts in which separate component sites participate in the reaction. Nevertheless, it has been suggested that carbonium ion-like intermediates are involved in alkane isomerization reactions on platinum,[30] and a specific mechanism has been proposed by Anderson and Avery.[31]

Dehydrocyclization is similar to isomerization, since it can also occur on the metal alone or by both the precious metal and the acid portion of the catalyst. Dautzenberg and Platteeuw[32] report that the dehydrocyclization of *n*-hexane to benzene occurs over a catalyst in which platinum is supported on a nonacidic alumina. It implies that the reaction proceeds over the metal component of the catalyst, since bifunctional catalysis with participation of acidic sites is then presumably eliminated. Gates et al.[8] proposed the networks to account for the observations made by Dautzenberg and Platteeuw.[32]

The following reaction networks illustrate the isomerization and dehydrocyclization of *n*-hexane and 2-methylpentane catalyzed by Pt on nonacidic alumina under reforming conditions:[8]

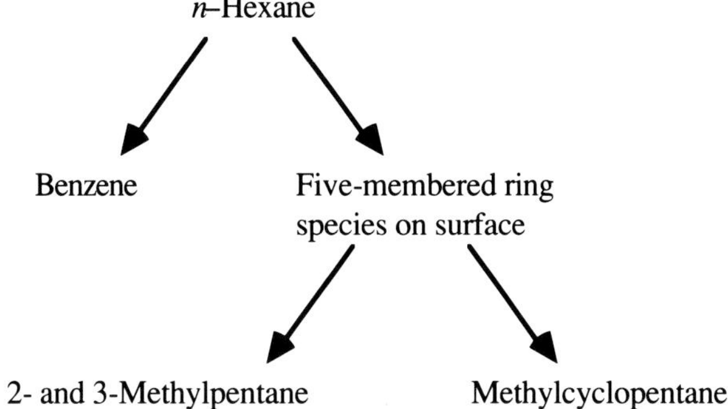

The reaction of branched paraffins containing only five carbon atoms in the chain involves the formation of a five-membered ring species, which then gives isomerized products, including *n*-hexane. The latter may form a six-membered ring, according to:

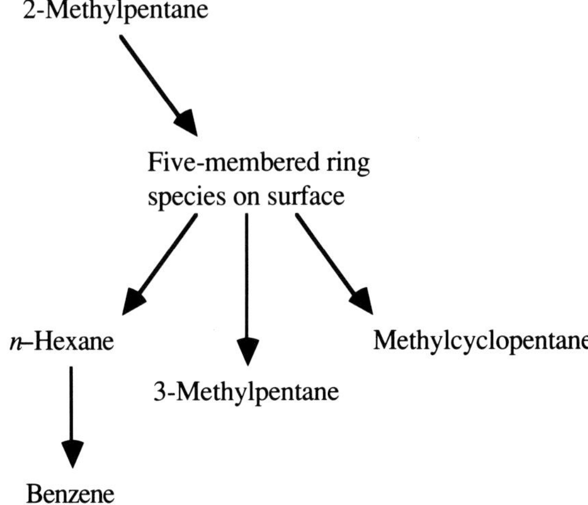

Reactions involving rupture of carbon-carbon bonds with accompanying hydrogenation (i.e., hydrocracking and hydrogenolysis) have been studied extensively in recent years by Sinfelt and co-workers.[33,34] Hydrogenolysis, a part of the overall reaction, requires only the metal component of the catalyst. The sequence of steps for hydrogenolysis appears to involve (1) chemisorption of hydrocarbon reactant with dissociation of carbon-hydrogen bonds and (2) rupture of carbon-carbon bonds in the unsaturated hydrocarbon residue formed in the chemisorption step. The species formed in the later step are then hydrogenated to form the products of the reaction. Hydrocracking, the other part of the reaction, involves the acidic site of the carrier. Sinfelt et al.[3,34] suggested that saturated hydrocarbons undergo hydrocracking by two different routes, one involving only the acidic sites of the carrier and the other involving a bifunctional action of these sites with metal sites. However, for some reason, Pt is not highly active for the hydrocracking reaction. In general, saturated paraffins predominate among the cracked products because the olefins formed are hydrogenated on the metal components.

It also became clear from the earlier studies that under the industrial conditions the surface of platinum is covered by sulfur, carbonaceous residues, coke, etc. to such an extent that most of the reactions are severely slowed, and only the simplest reaction of dehydrogenation can still proceed.[22] These facts lead to the conclusion that the most important function of Pt is to catalyze dehydrogenation. The sulfur effects and the mechanism of coke formation have been dealt with in detail in a subsequent section.

B. Characterization of Dispersed Pt on Al_2O_3

The catalyst selectivity in reforming is also dependent on dispersion, which suggests the importance of geometric effects due to variations in crystallite size alone, or alternatively, due to structural effects induced by the support at a very high degree of metal dispersion.

Gas chemisorption has been valuable for determining the surface areas of metals on supports. The degree of dispersion of catalytically active platinum on several different supporting materials has been a subject of a number of earlier investigations.[18] It was demonstrated that the selective chemisorption of gases (principally hydrogen, oxygen, and carbon monoxide) can provide a measure of the exposed metal surface area.

The research efforts carried out in the early 1950s were also directed toward defining the acid component characteristics. During this period, available procedures for measuring the acidity of silica-alumina and alumina catalysts were utilized to define the properties of supports for the platinum-reforming catalysts. These have been summarized in some of the previous reviews.[18,81] The following conclusions are drawn from these two reviews. The studies have shown that (1) thermal and hydrothermal treatment of the support not only causes a decrease in surface area but also a decrease in the measurable acidity, (2) it is difficult on a quantitative basis to compare the acidities of the various supports and interpret these in terms of their ability to carry out the cracking reactions of hydrocarbons, and (3) the alkali metals in general depress the acidity of most of these supports.

In spite of some disagreements among the various experimental measurements reported in the early years, there are two important points revealed in these studies. First, for freshly prepared Pt-Al_2O_3 catalysts with platinum contents in the range of those used industrially and which have not been subjected to severe heat treatment, it appears that virtually every platinum atom is available for chemisorption. It should be noted that this does not imply complete dispersion of the metal, since the method of measurement is not necessarily capable of discriminating between separated individual metal atoms and continuous monolayer 'islands' on the surface of the support. However, in regard to this point, several workers[35,36] have inferred that the platinum is not truly atomically dispersed, since, for example, as pointed out by Spenadel and Boudart,[35] it would be difficult to account for the rapid uptake of one hydrogen atom per platinum atom in the sample if a considerable portion of the platinum exists as discrete atoms capable of adsorbing only one hydrogen atom each. The stoichiometrics of chemisorption of oxygen and hydrogen and of the H_2-O_2 or O_2-H_2 titrations have been confirmed and accepted for Pt-Al_2O_3 catalysts.[37]

Second, after use under industrial conditions or after excessive heating in various atmospheres, the extent of gas adsorption decreased. In the limit of rather severe sintering of metal, quite satisfactory agreement was obtained between mean particle sizes as calculated from gas adsorption and from X-ray diffraction-line broadening.[35,38] Thus, it appears that the loss of platinum surface area is due primarily to the growth of platinum crystallites on the surface of the support at the expense of the more finely divided metal. A long debate still centers around the cause of sintering of platinum particles.

Somewhat later, electron microscopy studies provided independent evidence for the extremely high dispersion of the platinum.[38] Such studies have shown that the platinum exists as very small clusters of the order of 10 Å in size. TEM has also permitted direct observation of the sintering process in noble metal catalysis.[38] By measuring particle size as a function of sintering time, the change in particle size distribution and cluster growth can be determined.

In most of the reforming catalysts, the metal loading and the size of the platinum particles are so small that conventional surface characterization techniques have not been very successful. For example, XPS investigations have not been useful in determining the chemical state of Pt in alumina for two reasons. First, the Pt 4f features, which are sensitive to chemical state because they are high-lying core levels, are completely masked by the large Al 2p peak. Second, other Pt peaks, such as the 4d, are relatively insensitive to chemical state.

Recent measurements of EXAFS (extended X-ray absorption fine structure spectroscopy) have, however, been very valuable for very highly dispersed catalysts. From the EXAFS data,[40] values of average coordination number, interatomic distance, and disorder parameter were obtained for the platinum clusters. The average number of nearest-neighbor atoms about a platinum atom in a cluster was 7 for the Pt-Al_2O_3, significantly lower than the value of 12 for bulk platinum. This should be expected, since most of the platinum atoms in the clusters are surface atoms with lower coordination numbers than the atoms interior of a crystal. Exxon researchers concluded that the metal clusters behave as if they are more electron deficient than the bulk metal,[3,10] but others disagree.

Further progress in the characterization of dispersed metal catalysts has been made as a result of application of other special techniques, including Mössbauer spectroscopy[10] and Rutherford backscattering;[4] the efforts were, however, focused on the understanding of the role of the second metal in bimetallic clusters. They are therefore taken up with the characterization of bimetallic catalysts.

C. Effects of Crystallite Size on Reforming Reactions

The relationship between the particle size and their reactivity is one of the important questions in heterogeneous catalysis. Often, contradictory information has been obtained concerning the particle-size dependence of the reforming reactions. Normal hexane has been usually chosen as a test reaction since it is the simplest molecule which can undergo both the reactions, dehydrocyclization, and isomerization.

In the early 1960s, the French authors[42,43] noticed very large differences between the concentrated (5 to 20%) and dilute (0.2 to 1.5%) Pt-Al_2O_3 catalysts, as far as the product distributions were concerned; the differences were attributed to an effect of the metallic particle size. At a later time, Dautzenberg and Platteeuw[32,44] repeated the experiments of Barron et al.[42] (with 0.5 and 1% metal on the carrier), in hydrogenolysis of methylcyclopentane and isomerization of *n*-hexane or methylpentanes, and obtained exactly the same product distribution as the French workers. With this limited study they claimed, however, that the very large difference of selectivity observed by Barron et al. was related to a difference of chlorine content due to the mode of preparation, and not to a particle size effect. Subsequently, the French group[45] showed that in no case the chlorine content, and more generally any dual function mechanism, may account for the differences observed between the concentrated and dispersed platinum alumina catalysts.

Similarly, working on 15 to 80 Å Pt particles supported on alumina or silica, Lankhorst et al.[46] have detected no significant particle size dependence on *n*-hexane isomerization and cyclization. In contrast, Santacessaria et al.[47] and Anderson and Shimoyama[48] observed increased activity for *n*-hexane isomerization, hydrogenolysis, and C_5-cyclization on very small particles (<10 Å to >100 Å). The latter experiments were carried out on thin films in order to avoid the complication due to possible surface contamination. The particle-size dependence was explained in terms of a change of mechanism from cyclic to bond shift as the particle size was increased.[49,50]

Results of Davis et al.,[51a] where experiments were conducted with *n*-hexane on platinum single-crystal surfaces, showed the most significant structure sensitivity for the aromatization to benzene. The face (111) showed the maximum increase in the rate and selectivity. On the other hand, isomerization of light alkanes, *n*- and i-butanes, displayed a significant structure sensitivity, high concentrations of (100) require that the EXAFS be measured over an interval which is greater than about 16 $Å^{-1}$ (1/0.6 Å; the precision of any frequency measurement is given by the inverse of the length of the interval over which the oscillations are measured) in electron momentum or 1000 eV in electron energy. Kelley and co-workers,[85] utilizing XPS and XAS, further showed that Pt^0 and Re^{4+} are the dominant species after reduction, and also suggested that Re-Cl interactions are important.

The effect of sulfur has also been studied since it is commonly added to catalysts of this type in reforming practice.[9] Although immediately recognized by industry,[114] the beneficial effects of sulfur on the activity and selectivity of Pt-Re-Al_2O_3 and Pt-Al_2O_3 catalysts were demonstrated by Menon and Prasad.[115] Since then, the effects of sulfur adsorption on the catalytic properties of metal have been discussed both in terms of geometric and electronic effects.[116] Most authors, however, invoke geometric effects to describe changes in activity and selectivity by clustering.[22]

The studies of Shell workers (utilizing chemisorption, IR, and XPS),[108,109] have shown that rhenium atoms covered with the chemisorbed sulfur atoms divide the metal surface of platinum-containing catalysts into ensembles consisting of a microfacets appearing to be necessary for high catalytic activity in butane isomerization.[52] These authors attributed the difference in structure sensitivity behavior between the isomerization rates of *n*-hexane and the butanes to mechanistic differences; a cyclic mechanism is suggested to be operative in the hexane case, whereas light alkanes probably react via a bond-shift mechanism.

Their recent platinum single-crystal surface studies[51b] employing LEED and AES showed that the aromatization of *n*-hexane to benzene displayed unique structure sensitivity in which the rates for this important reforming reaction were maximized on platinum surface with terrace structure. The rates of competing isomerization, cyclization, and hydrogenolysis reactions displayed little dependence on surface structure. Catalyst deactivation resulted from the formation of disordered carbonaceous deposits on the platinum surfaces, whose primary role was that of a nonselective poison.

In summary, the investigations carried out during the 1950s have provided a fairly good picture of the effect of method of catalyst preparation, nature of the acidic component, catalyst activation procedures, and thermal and hydrothermal conditions of the finished catalyst. In the 1960s, the results of various investigations provided further information on the effects of these factors on catalyst activity, selectivity, and stability. In recent years, research activities were directed toward determining the nature of the platinum in reforming catalysts, mainly by utilizing special techniques, with the aim to obtain a more fundamental understanding of the catalyst and how it functioned. This provided the firm foundation for the concept of bifunctional catalysis, in which two different catalytic functions are present in a single catalyst. These advances also stimulated research leading to significant progress in understanding the role of second metal in the bimetallic reforming catalysts. While the increased catalytic activity, selectivity, and stability all resulted from a succession of catalyst modifications, these functions were only sometimes recognized before they were built into the catalyst.

IV. BIMETALLIC CATALYSTS: CHARACTERIZATION

Modern reforming catalysts are bimetallics. The three most extensively studied bimetallic reforming catalysts are Pt-Re, Pt-Ir, and Pt-Sn, all supported on alumina. Concerted special techniques are needed to see the individual platinum and the second metal particles, and/or bimetallic clusters in these catalysts. In the last 2 decades, researchers have made considerable progress in understanding the bimetallic catalysts. Nevertheless, how the modifier metal functions and how it is disposed in the catalyst remains unclear.

In recent years, the utilities of X-ray absorption fine structure (EXAFS) and Mössbauer spectroscopy, complemented by Rutherford backscattering spectrometry, as the probes for the study of the highly dispersed bimetallic catalysts have been clearly demonstrated. EXAFS has been particularly useful for the characterization of bimetallic clusters. Of the studies published so far on the bimetallic cluster catalysis, those by Sinfelt (Exxon) and his collaborators, Via (Exxon), and Lytle (Boeing) are particularly comprehensive.[2,10,51-58] The following discussion proceeds with the characterization of Pt-Re, Pt-Ir, and Pt-Sn reforming catalysts mainly by the use of these special techniques. Some selected experimentations are also described.

Bimetallic catalysts of Pt and Ir, Re, or Sn can be prepared by coimpregnating (incipient wetness) a carrier such as alumina with an aqueous solution of chloroplatinic acid and chloroiridic acid or Re_2O_7 or $SnCl_2$, respectively. After the impregnated carrier is dried and calcined (Pt-Re and Pt-Sn, approximately 700 K; mild conditions for Pt-Ir, 525 to 540 K), subsequent treatment in flowing hydrogen at temperatures (approximately 700 K) leads to formation of bimetallic clusters. For comparison, individual metal catalysts can also be prepared under the same conditions except for Re, where the calcination is not performed. As a guideline, the clusters of interest have sizes of the order of 100 Å and smaller. Figure 5 illustrates Pt-Ir clusters which appear as black dots in an electron micrograph of a catalyst in which the clusters are dispersed throughout aluminum oxide.[56] The clusters measure about 10 Å.

A. Pt-Re System

Platinum-rhenium catalysts are particularly effective for the selective conversion of cycloalkanes into aromatic hydrocarbons. The controversy centers around the problem whether Re and Pt are separate entities or whether Re is clustered with platinum in reduced catalyst. It has also been suggested that perhaps Re reacts much more strongly with alumina at the few tenths weight-percent loadings typical of industrial catalysts.

The conclusion of Johnson and Leroy,[61] and Johnson[62] using chemisorption, IR, and XRD studies indicated that complete reduction of Pt from the +4 oxidation to metal and/or Re from +7 to +4 state, and the study of XRD data on the metal residue from the leached catalysts showed no evidence for the presence of Re metal or Pt-Re alloy. In contrast, chemisorption studies by Webb[63] on $Re\text{-}Al_2O_3$ catalysts indicated that Re was completely reduced from the +7 state to the metal.

Later experiments using chemisorption and IR,[64-70] and TPR[71] were consistent with the results of Webb[63] in showing a change in oxidation state of Re from +7 to 0. These workers also indicated that the properties of $Pt\text{-}Re\text{-}Al_2O_3$ catalysts depended on the method of preparation. In the case of catalysts that were simply dried in air at 383 K after impregnation of alumina with H_2PtCl_6 and Re_2O_7, it was concluded that a Pt-Re alloy formed on reduction. However, from the catalysts that were calcined in air at 773 K, prior to reduction in hydrogen, it was concluded that the platinum exhibited much less interaction with rhenium.[72-75] The valence-state problem seems natural for XPS, since the binding-energy shifts for the intense 4f peak are large and well known.[76] Using XPS, Biloen et al.[77] concluded the formation of a Pt-Re alloy (on silica).

More direct evidence of an interaction between Pt and Re has been obtained using TPR.[78] This work demonstrates that Pt and Re on $Pt\text{-}Re\text{-}Al_2O_3$ catalysts do not behave as the sum of individual components but suggests a bimetallic cluster, alloy, or an interaction through

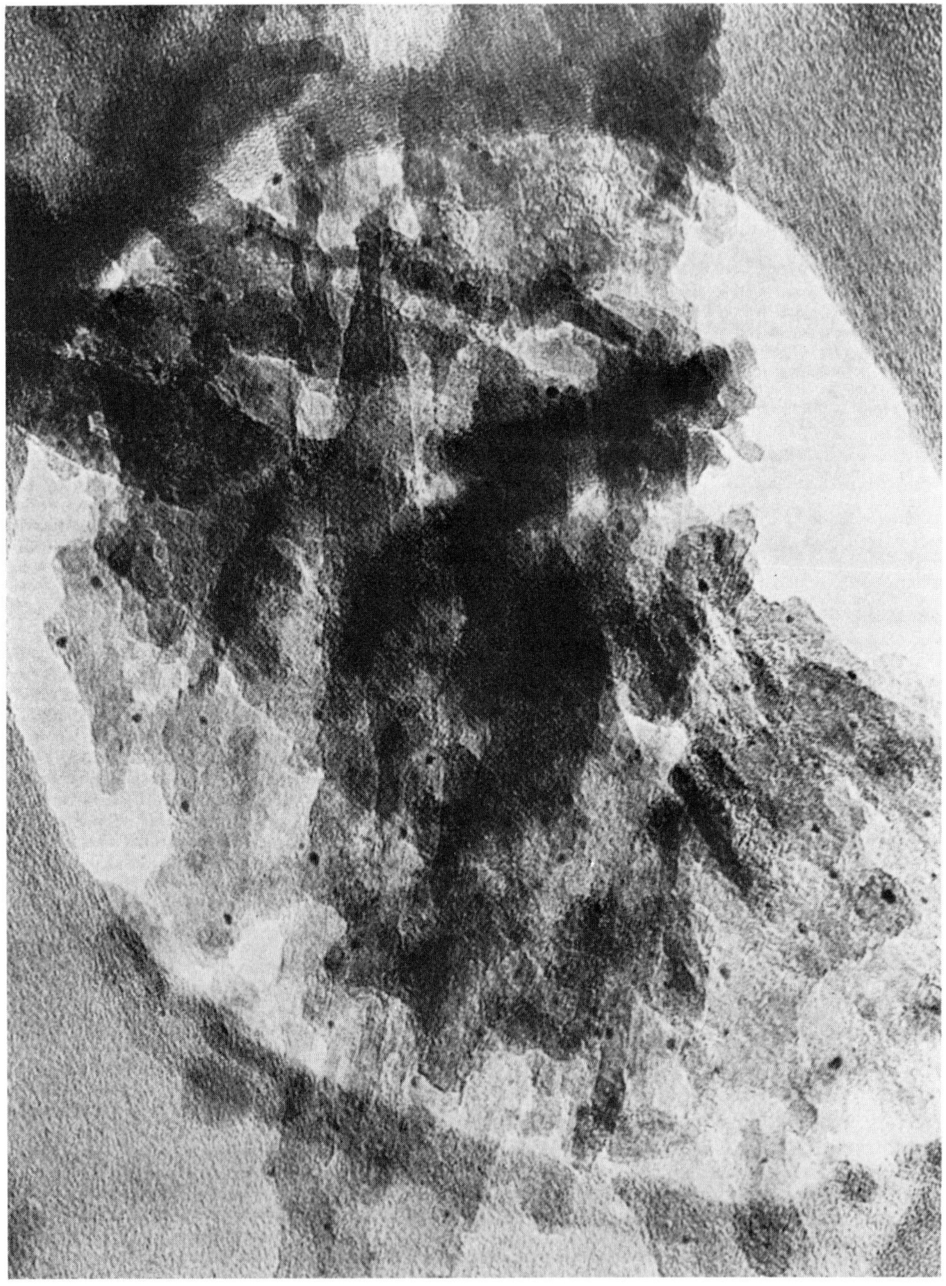

FIGURE 5. Electron micrograph of Pt-Ir clusters dispersed in Al_2O_3. The clusters measure about 10 Å. (From Sinfelt, J. H., *Sci. Am.*, 253, 90, 1985. With permission.)

a support modification. In contrast, IR study indicated no evidence for the formation of Pt-Re alloys.[79] Also, the catalytic evidence favored separate dispersion: a mixed bed of Pt and Re on separate pellets performed similarly to coimpregnated pellets with no evidence of Re transfer.[80] Surface analysis by ISS and microanalysis by energy-dispersive X-ray spectroscopy in scanning transmission electron microscope studies indicate that Re is not significantly associated with Pt, but rather is widely dispersed on the support surface.[5]

Table 4
ENERGIES OF X-RAY ABSORPTION EDGES FOR Pt AND Re[86]

Edge	Energy (keV)
Re L_I	12.525
Re L_{II}	11.957
Re L_{III}	10.534
Pt L_I	13.892
Pt L_{II}	13.273
Pt L_{III}	11.564

The small metal particles typical of industrial reforming catalysts give X-ray diffraction features so broad that they are not useful for lattice parameter determination. Large particles are obtained when such catalysts are thermally sintered, and they show the lattice parameter charges expected for alloy formation.[65] Efforts to obtain electron diffraction from Pt-Re-Al_2O_3 catalysts have been unsuccessful.[83] Pure rhenium-supported catalyst showed no structural details.[4]

While the Re EXAFS offers enough information to identify neighboring atoms as oxygens for Pt-Re system,[84,85] the Pt EXAFS has been of little value since the maximum range over which EXAFS following any Pt edge is free of other fine structure is 393 eV (Table 4[86]). For EXAFS studies, the extended fine structure associated with an absorption edge should have about 1000 eV beyond the edge in order to be free of other absorption edges; otherwise, there will be an overlap of the EXAFS associated with each edge and the data will be difficult to interpret (Figure 6). The terseness of this range limits the interpretation of Pt EXAFS since to estimate the frequency of EXAFS oscillations to a precision better than about 0.06 Å requires that the EXAFS be measured over an interval which is greater than about 16 $Å^{-1}$ (1/.06 Å; the precision of any frequency measurement is given by the inverse of the length of the interval over which the oscillations are measured) in electron momentum or 1,000 eV in electron energy. Kelley and co-workers,[85] utilizing XPS and XAS, further showed that Pt^0 and Re^{4+} are the dominant species after reduction, and also suggested that Re-Cl interactions are important.

The effect of sulfur has also been studied since it is commonly added to catalysts of this type in reforming practice.[9] Although immediately recognized by industry,[114] the beneficial effects of sulfur on the activity and selectivity of Pt-Re-Al_2O_3 and Pt-Al_2O_3 catalysts were demonstrated by Menon and Prasad.[115] Since then, the effects of sulfur adsorption on the catalytic properties of metal have been discussed both in terms of geometric and electronic effects.[116] Most authors, however, invoke geometric effects to describe changes in activity and selectivity by clustering.[22]

The studies of Shell workers (utilizing chemisorption, IR, and XPS,[108,109] have shown that rhenium atoms covered with the chemisorbed sulfur atoms divide the metal surface of platinum-containing catalysts into ensembles consisting of a small number of contiguous atoms. Thus, the Shell researchers concluded that the Pt-Re bonds were present at the surface in platinum-rhenium catalysts, whether or not chemisorbed sulfur is present.

A more recent application of proton-induced X-ray emission analysis (PIXE) and Rutherford backscattering spectrometry (RBS) (together with electron microscopy and chemisorption studies) has provided a significant insight into the near-surface elemental composition.[4,110] In considering the nature of Pt-Re catalysts, we begin with a comparison of the chemisorption properties of alumina-supported rhenium, platinum, and platinum-rhenium catalysts. Typical data on the chemisorption of oxygen are shown in Figure 7. The

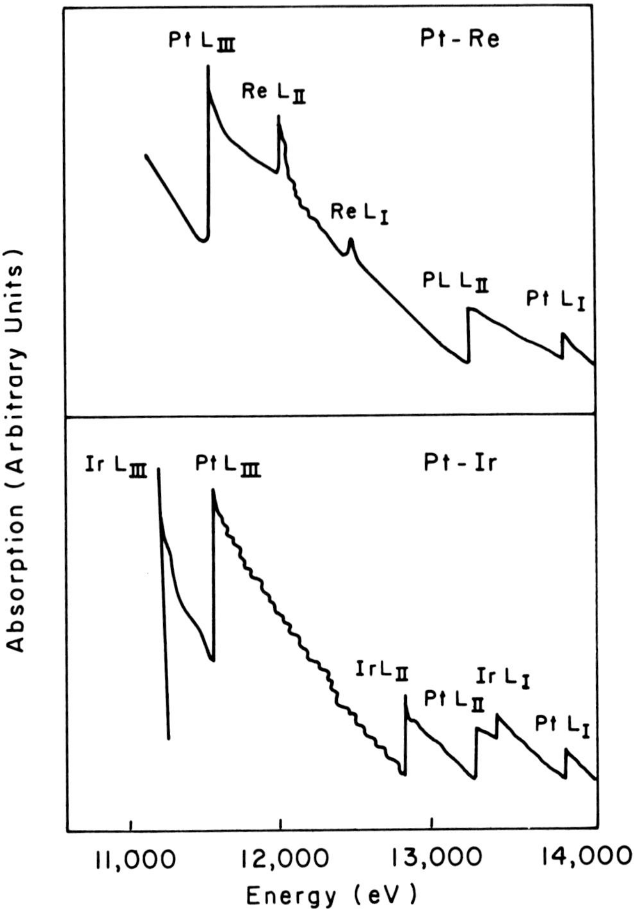

FIGURE 6. X-ray absorption of Pt-Re and Pt-Ir clusters. (From Sinfelt, J. H., Via, G. H., and Lytle, F. W., *J. Chem. Phys.*, 76, 2779, 1982 and Onuferko, J., Short, D. R., and Kelley, M. J., *Appl. Surf. Sci.*, 19, 227, 1984. With permission.)

extrapolation of the quasilinear part of the oxygen adsorption isotherm to zero pressure gives a value for the amount of oxygen chemisorbed at room temperature.

Isotherms A and B show (Figure 7) the chemisorption measurements on Pt-Al_2O_3 and Re-Al_2O_3 catalysts, respectively. Two isotherms are shown for Pt-Re-Al_2O_3 catalyst. Isotherm C represents the total oxygen chemisorbed by Pt and Re in Pt-Re-Al_2O_3 catalyst, whereas isotherm D shows the chemisorption of oxygen by Pt in Pt-Re-Al_2O_3 catalyst. Oxygen chemisorption on Pt in a Pt-Re-Al_2O_3 catalyst was calculated from oxygen uptake of Pt-H after reduction of Pt-Re-Al_2O_3 catalyst under hydrogen at room temperature. Reduction of Re-O to Re-H at this room temperature was assumed to be negligible. The difference between oxygen uptake calculated from isotherms C and D is a measure of the amount of oxygen chemisorbed by Re only in Pt-Re-Al_2O_3 catalyst.

Table 5 summarizes their chemisorption data for Pt-Al_2O_3, Re-Al_2O_3, and Pt-Re-Al_2O_3 catalysts. Dispersion of Pt and Re was calculated from the stoichiometry of gas titration as given by Menon et al.[64] The presence of Re seems to improve the dispersion of the bimetallic catalyst. While this chemisorption behavior is consistent with the possibility that bimetallic clusters of platinum and rhenium are present, they do not rule out the possibility that the platinum and rhenium are present in small fractions as unclustered entities in the catalyst.

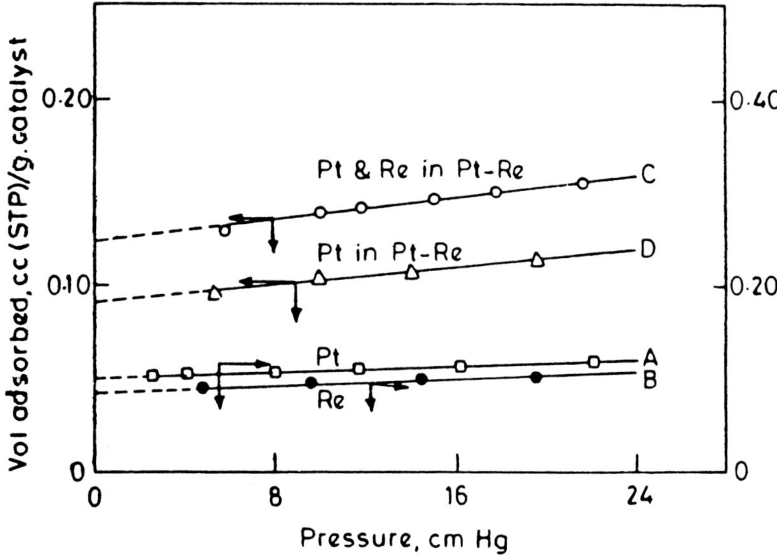

FIGURE 7. Typical isotherms at room temperature for chemisorption of oxygen on the alumina-supported Pt, Re, and Pt-Re catalysts. A,B: chemisorption of oxygen on Pt and Re catalysts, respectively. C,D: oxygen chemisorbed by Pt and Re, and Pt in Pt-Re catalysts. (From Jothimurugesan, K., Nayak, A. K., Mehta, G. K., Rai, K. N., Bhatia, S., and Srivastava, R. D., *AIChE J.*, 31, 1997, 1985. Reproduced by permission of the American Institute of Chemical Engineers.)

Table 5
SUMMARY OF CHEMISORPTION DATA ON Pt-Al_2O_3, Re-Al_2O_3 AND Pt-Re-Al_2O_3 CATALYSTS

Catalyst	Calcination (Temp., K[a])	Reduction (Temp., K[b])	% dispersion	Metal area (m^2/g)
0.3% Pt	723	773	39.1	0.29
0.3% Re	373[c]	773	31.3	0.23
0.3% Pt, 0.3% Re	723	773	48.0	0.35

[a] Calcination conducted in air for a period of 5 hr.
[b] Reduction conducted in flowing hydrogen for 2 hr.
[c] Drying temperature.

From Jothimurugesan, K., Nayak, A. K., Mehta, G. K., Rai, K. N., Bhatia, S., and Srivastava, R. D., *AIChE J.*, 31, 1997, 1985. Reproduced by permission of the American Institute of Chemical Engineers.

Figure 8 shows the RBS spectra of the calcined and reduced Pt-Re catalyst.[4] A striking feature is a sharp peak of the Pt-Re in the reduced sample. Figure 9A and B are RBS spectra of the sulfided and deactivated (accelerated deactivation by using the catalyst for 24 hr in the dehydrogenation of methylcyclohexane without hydrogen) catalyst.[110] The sulfided catalyst showed behavior similar to that of the reduced one. It was, of course, not possible to distinguish Pt and Re separately because of the limitation of the proximity of their mass numbers. However, RBS results showed that in the deactivated sample the surface concentration of Pt-Re decreased from 15 $\times$ 10^8 to 3 $\times$ 10^8 atoms/cm^3. PIXE studies (Figure 6, Chapter 2) revealed that, in the case of calcined catalyst, the distributions of Pt and Re were identical, however, for reduced catalyst, Re concentration seemed to predominate over Pt.

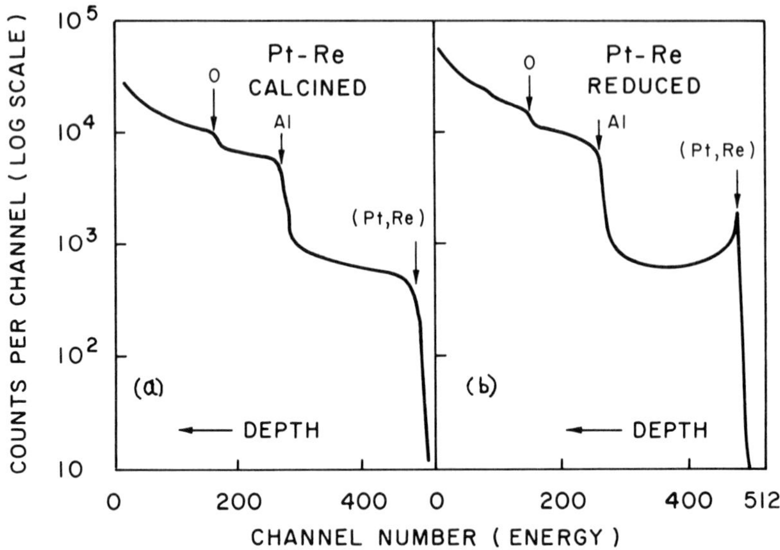

FIGURE 8. Backscattering spectra of Pt-Re catalyst (a) calcined, (b) reduced. (From Jothimurugesan, K., Nayak, A. K., Mehta, G. K., Rai, K. N., Bhatia, S., and Srivastava, R. D., *AIChE J.*, 31, 1997, 1985. Reproduced by permission of the American Institute of Chemical Engineers.)

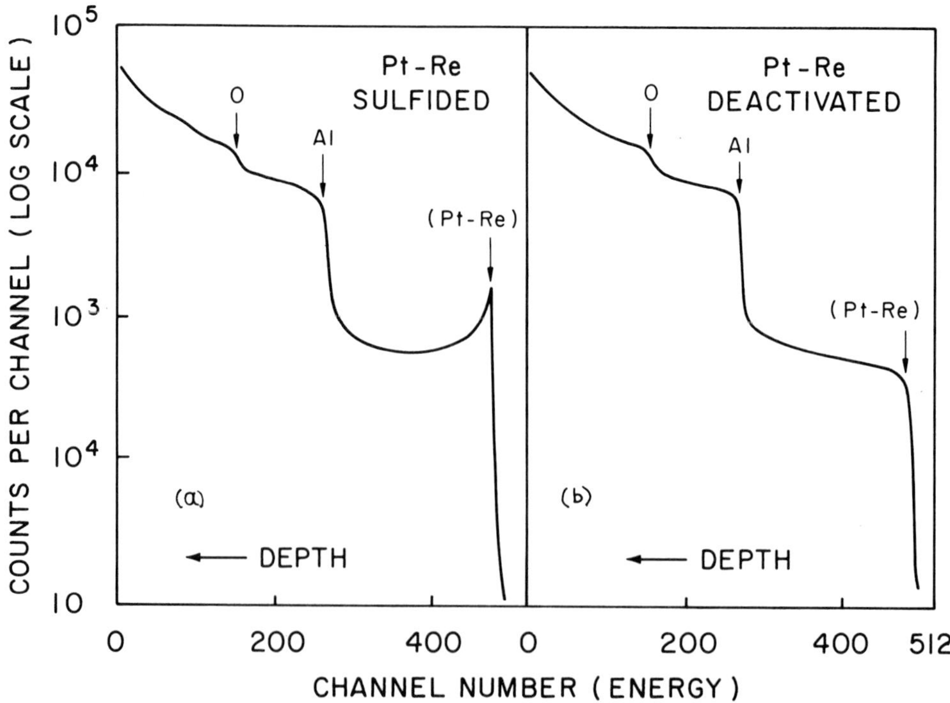

FIGURE 9. RBS spectra of Pt-Re-Al_2O_3 catalyst (a) sulfided, (b) deactivated. (Reprinted with permission from Srivastava, R. D., Prasad, N. S., and Pal, A. K., *Chem. Eng. Sci.*, 41, 719, 1986, Pergamon Press, Ltd.)

The results of TEM studies[4,110] showed that Pt-Re-supported catalyst exhibited dispersed particulate morphology with a ring diffraction pattern that had the same crystal structure as pure platinum catalyst. This observation is expected because rhenium forms a substitutional solid with Pt, without altering the Pt structure. The solid solubility of Re in Pt up to 41% has been reported.

From the foregoing results, it appears that Pt-Re has a lot to do with the microstructures that might be present in the catalyst, i.e., alloy, separate metal particles, ionic Re, etc. These all relate to different hypotheses of how the catalysts function, e.g., subdividing the active metal surface into ensembles too small to permit coke formation. At this stage of discussion, it is, however, clear that the chemisorbed sulfur atoms are present on surface rhenium atoms, and that the latter remain bonded to platinum atoms in the surface region.

B. Pt-Ir System

This system has received increased attention primarily by the researchers at Exxon. Under reforming conditions, Ir is more active than Pt for aromatization, but also unfortunately for cracking. The known hydrogenolysis activity of iridium has been claimed to be eliminated or at least controlled.[87] Considerable effort was devoted in establishing that Pt-Ir catalysts are indeed bimetallic clusters. TPR study by Wagstaff and Prins[88] shows that both elements in alumina-supported Pt-Ir are reduced to metal. Since virtually all the metal atoms are surface atoms, Mössbauer-effect studies and EXAFS investigations have been found to be very useful. We again begin with the chemisorption study for this system.

Sinfelt et al.[10] used hydrogen chemisorption in order to evaluate the variation of metal dispersion with variation of the total metal content. The quantity H/M, which represents the ratio of the number of hydrogen atoms adsorbed to the number of metal atoms (platinum and iridium) in the catalyst, was taken as a measure of the dispersion. This value approached a limiting value near unity as the metal concentration was decreased below 1 wt%. Their electron microscopy data showed the average diameter of the metal clusters to be of the order of 10 Å or lower.

XRD has also been utilized to study this sytem, and has been reported to be useful with metal dispersions as high as 0.6.[89] XRD data on a silica-supported platinum-iridium catalyst containing 10 wt% Pt and 10 wt% Ir revealed the lattice parameter of the platinum-iridium clusters as 3.875 Å. These authors also reported an average platinum-iridium cluster size of 49 Å, as calculated from the line width measurement by Scherer formula. The metal dispersion of the catalyst as determined by hydrogen chemisorption technique was 0.24.

An X-ray absorption spectrum at 100 K in the region of the L absorption edges of iridium and platinum is also included in Figure 6 for a Pt-Ir catalyst containing 10 wt% each of the individual metals.[51] The L_{III} absorption edges of iridium and platinum are 348.5 eV apart in energy. Again there is overlap of the EXAFS associated with the L_{III} of iridium and platinum. This is because the extended fine structure associated with the L_{III} edge of Ir is observable to energies of 1200 to 1300 eV beyond the edge. However, these authors were successful in separating the iridium EXAFS from the platinum EXAFS in the region of overlap. This was accomplished by obtaining EXAFS data on Pt-Ir materials with known structural parameters; the materials included a physical mixture of platinum and iridium and a bulk alloy of Pt and Ir.

From results on interatomic distances derived from analysis of EXAFS data, Sinfelt and co-workers[52] inferred that the clusters were nonhomogeneous. They based their argument on simple logic that if the clusters were truly homogeneous, the interatomic distance derived from the platinum EXAFS should be identical to that derived from the iridium EXAFS. These authors found, in general, that the distances were not equal. Also, based on surface energy considerations (platinum has a lower surface energy than iridium), they concluded that the platinum concentrates at the surface or boundary of the clusters. They further

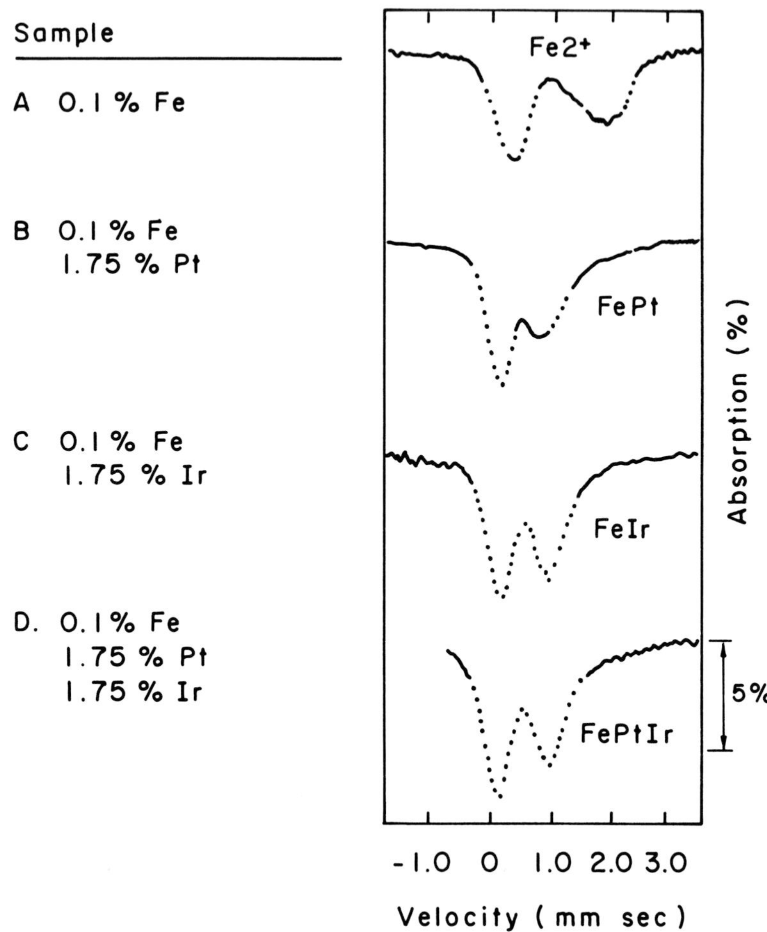

FIGURE 10. Mössbauer spectra at 298 K on alumina-supported platinum, iridium, and platinum-iridium catalysts. (From Garten, R. H. and Sinfelt, J. H., *J. Catal.*, 62, 127, 1980. With permission.)

suggested that, in the case of very highly dispersed platinum-iridium clusters on alumina, the clusters may very well have 'raft-like' two-dimensional structures.[55] The perimeter of a central iridium or iridium-rich raft would then be concentrated with platinum.

The Mössbauer spectra of iron doped into Pt-Ir-Al_2O_3 resembles that seen for low-dispersion alumina-supported Pt-Ir, where cluster formation has been established by XRD.[55,87] The Mössbauer spectra are shown in Figure 10 for alumina-supported platinum, iridium, and bimetallic platinum-iridium catalysts containing ^{57}Fe samples (sample B, C, and D, respectively).[55] A is a reference material containing only the enriched Fe on alumina. All of the catalysts had metal dispersion in the range 0.7 to 1. The two-line (quadrupole split, QS) spectrum for sample A with the corresponding Mössbauer parameters indicated that the iron in the sample is reduced only to the ferrous state with hydrogen at 773 K. In sample B, incorporation of iron atoms into the platinum clusters was indicated by the agreement of the isomer shift (IS) for iron in the catalyst with that for iron in a PtFe bulk alloy of similar composition. Sample C, containing Ir and Fe, also indicated the incorporation of the iron into the iridium clusters. Finally, the Mössbauer parameters for sample D differed from those for samples B and C, which indicated that iron atoms in sample D are not associated exclusively with either the platinum or the iridium in the catalyst. Thus, the Mössbauer data

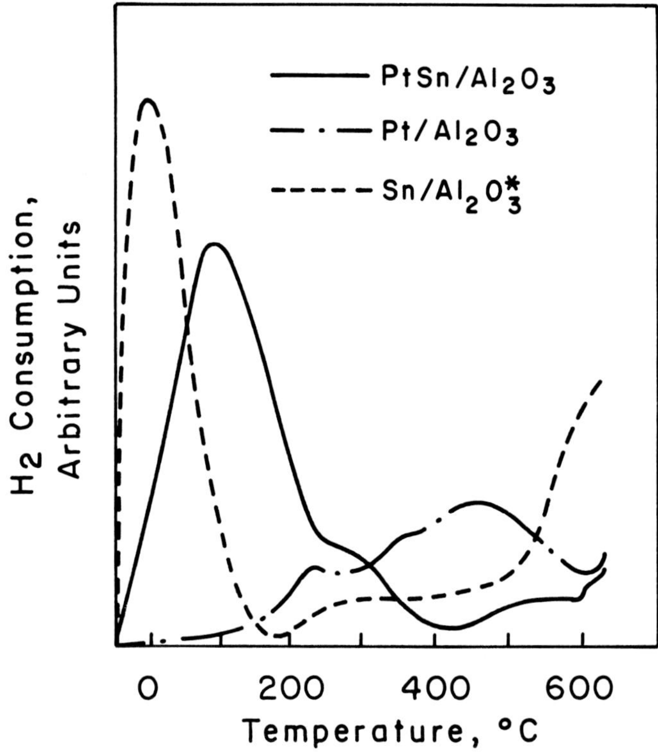

FIGURE 11. TPR profiles after oxidation at 453 K. (From Dautzenberg, F. M., Helle, J. N., Biloen, P., and Sachtler, W. M. H., *J. Catal.*, 63, 119, 1980. With permission.)

also support the view that platinum and iridium atoms form bimetallic clusters, rather than forming separately supported monometallic clusters consisting exclusively of atoms of one or the other of the two metals.

C. Pt-Sn System

Another catalyst system that may also be considered in the bimetallic cluster category is Pt-Sn, which represents another type of system in the sense that Group IVA metallic element (Sn) is incorporated with the group VIII metal component (Pt). The literature describes five phases in the PtSn system: Pt_3Sn, PtSn, Pt_2Sn_3, $PtSn_2$, and $PtSn_4$. Various investigators[90-96] have shown that the addition of tin to platinum-alumina catalyst improves the activity and selectivity of the catalyst.

Early AES and XPS results of Bouwman and co-workers[98,99] in conjunction with chemisorption, showed how the surface composition of alloys depends on the gas phase. These studies revealed surface enrichment of PtSn and Pt_3Sn with tin. Moreover, evidence was also obtained for the formation of SnO_2 complexes on the exposed surface of Pt_3Sn. Berndt et al.[100] found evidence of an interaction between platinum and tin, but could not differentiate between PtSn and PtSnO type of clusters. Their chemisorption study showed a decrease in the adsorption capacity of Pt in the presence of Sn which strongly corroborated the possibility of PtSn cluster formation. TPR results of Dautzenberg et al.[92] showed that the profile obtained for the Pt-Sn-Al_2O_3 sample differed considerably than that from superimposing the separate profiles for Pt-Al_2O_3 and Sn-Al_2O_3 (Figure 11). Thus, it was concluded that in the reduced-supported Pt-Sn catalysts, bimetallic clusters were present.

Later work by Sexton et al.,[96] Burch,[101] Leiske and Volter,[102] and Adkins and Davis[103] aimed at analyzing the valence state of the tin atoms in working with Pt-Sn catalysis. Chemisorption studies on platinum-tin catalyst performed by Burch[98] showed that platinum catalyzed the reduction of tin, whereas tin increased the dispersion of Pt. However, the study revealed that the average oxidation state of tin after reduction was Sn(II) which could not be further reduced even by prolonged reduction and that proper alloys of Pt-Sn were absent in the catalyst. Similar results were obtained earlier by Muller and co-workers.[104] Lieske and Volter's[102] study on Pt-Sn-Al_2O_3 catalyst using TPR and adsorption of hydrogen and oxygen, revealed that Sn(IV) was mostly converted to Sn(O). Further, they found that, depending upon the tin content, Sn(O) formed a cluster with platinum of appropriate composition. In contrast, Adkins and David,[103] using ESCA to characterize reduced Pt-Sn-Al_2O_3 catalysts, found evidence of metallic platinum whereas metallic tin and SnO_2 were not detected.

The use of different impregnation techniques led to different adsorbed species in the Pt-Sn/Al_2O_3 catalyst precursors. Diffuse reflectance spectroscopy and TPR[105] showed that when the catalyst was prepared by coimpregnation or two-step impregnation (Pt deposition, drying, and Sn deposition), a Pt-Sn complex was adsorbed on the alumina surface. However, the catalyst prepared by two-step impregnation but in reverse sequence, showed only Pt(IV) and Sn(IV) species on alumina.

A number of Mössbauer spectroscopic studies have been conducted on Pt-Sn catalyst systems.[106,107] More recently, Srivastava and co-workers[111] carried a systematic study to determine the function of tin using an *in situ* Mössbauer spectroscopy, which has emerged as a potential tool for characterization of fine clusters. Their Mössbauer results, combined with TEM, XRD, and RBS results, completed the mechanistic picture of the catalytic action of Pt-Sn-Al_2O_3 preparations which have, of course, been subjected to a series of catalytic tests. The catalytic tests, along with RBS results, are described in Section V.

The spectra obtained for a sample, in which tin and platinum were added sequentially, showed that Fe^{+2} state grew at the cost of Fe^{+3} state. This was attributed to a dilution effect of the platinum atoms into small ensemble or due to migration of the platinum atoms to the tin atoms at the Fe^{+2} sites, forming bimetallic clusters. TEM and XRD also showed the formation of bimetallic clusters of intermetallic compound PtSn along with a small quantity of SnO.

V. REFORMING WITH BIMETALLIC CATALYSTS

As pointed out earlier in the text, the basic advantages of bimetallic reforming catalysts (Pt-Ir, Pt-Re, and Pt-Sn) are that they are much more stable and show higher selectivity for the formation of C_5+ reformate than the conventional platinum catalyst. Table 6 shows a typical pilot plant life test data for the higher and better activity maintenance of a platinum-iridium on alumina catalyst relative to one of the conventional high platinum on alumina catalysts.[112] The research octane number of the C_5+ liquid reformate reflects the activity of the catalyst. These results indicated that the octane number was substantially higher for the platinum-iridium catalyst than for the platinum catalyst, and showed the decline of the octane number lower for the platinum-iridium catalyst, and indication of its better activity maintenance.

Similarly, data on the temperature required to maintain 100 clear research octane number as a function of time on stream for the three commercial catalysts (0.3% Pt; 0.3% Pt-0.3%Re; 0.3% Pt-0.3% Ir) are given in Table 7.[87] This also shows that Pt-Ir requires a lower temperature to produce the desired octane number reformate. These aspects, along with the yield of C_5+ reformate, are frequently used for considerations in the choice of a catalyst.

Table 6
COMPARISON OF ALUMINA-SUPPORTED PLATINUM-IRIDIUM AND PLATINUM CATALYSTS

	Research octane number	
Hr on stream	**Pt(0.6%)**	**Pt(0.3%)-Ir(0.3%)**
100	99	103
200	97	102
300	96	102
400	95	102
500	94	102

Note: Test conditions: pressure: 14.6 atm; temp.: 760 K; feed: Venezuelan naphtha (323 to 473 K boiling range).

From Carter, J. L., McVicker, G. B., Weissman, W., Kmak, W. S., and Sinfelt, J. H., *Appl. Catal.*, 3, 327, 1982. With permission.

Table 7
DATA ON THE REFORMING OF A 372 TO 444 K BOILING-RANGE NAPHTHA TO 100 RESEARCH CLEAR OCTANE NUMBER PRODUCED OVER ALUMINA-SUPPORTED PLATINUM, PLATINUM-RHENIUM, AND PLATINUM-IRIDIUM CATALYSTS AT 14.3 ATM PRESSURE[87]

	Temperature (K)		
Hr on stream	**Pt**	**Pt-Re**	**Pt-Ir**
250	795	778	771
500	803	781	773
750	—	786	773
1000	—	788	773

It is well known that various types of hydrocarbon reactions contribute to the end results in the case of catalytic naphtha reforming. During this reforming process, desired reactions (paraffin isomerization and dehydrocyclization, and naphthene dehydrogenation and dehydroisomerization) and undesired reactions (hydrocracking to gases and coke deposition) are produced. The main property of the carbonaceous deposit is to poison the surface nonselectively, simply by hindering the access of reactants to the metal sites. However, the deposit may play an important role of exchanging hydrogen with reacting surface species, which in turn may alter the selectivity at high temperatures. The great advantage of the sulfurized catalyst is the stability, the catalyst maintaining a higher activity for longer time-on-streams. It has also been reported that the metallic function controls the aromatization reaction and is deactivated first.[132]

Consequently, it is important to consider the reforming properties of Pt-Re, Pt-Ir, and Pt-Sn catalysts along with their deactivation by deposition of coke in more detail. This section, therefore, proceeds with some information obtained from studies on the reforming of pure hydrocarbons and related deactivation studies over these catalysts which forms a basis in understanding of the way these catalysts function in reforming, followed by a summary of the results of some naphtha reforming runs.

A. Reforming with Pure Hydrocarbons and Deactivation Studies

In this section we consider both the influence of metals on the conversion patterns of naphthenic and paraffinic hydrocarbons and results of kinetic investigations of reactions of selected hydrocarbons catalyzed by bimetallic reforming catalysts, which are quite important in the reforming of naphtha feedstocks. As pointed out previously, the main reactions during naphtha reforming are dehydrocyclization, hydrocracking, and isomerization of paraffins and dehydroisomerization, hydrocracking, and isomerization of paraffins and dehydroisomerization of naphthenes. It is generally accepted that hydrocracking reactions are controlled by controlling the acidity of the acid components. Conversion of cyclohexane, methylcyclopentane (MCP), and methylcylohexane (MCH) to aromatics has often served as a model reaction of naphthenes, while *n*-heptane or *n*-hexane is chosen to model paraffin reactions. Some recent investigations on these reactions are summarized below.

The success of modern Pt-Ir catalysts can be viewed as using Pt to greatly reduce the hydrogenolysis activity of Ir while maintaining that for aromatization. Exxon researchers[112,113] performed extensive studies of *n*-heptane, methylcyclopentane, and cyclohexane conversion to provide insight into the advantages of Pt-Ir and Pt-Re catalysts over catalysts containing only Pt. The data were obtained at approximately 773 K and 14.6 atm using a 5:1 mole ratio of hydrogen to hydrocarbon.[112] The catalysts were presulfided, and also contained 0.9 wt% chlorine as charged.

The rates of *n*-heptane dehydrocyclization, r_1 (r_1 refers to the rate of production of toluene and C_7 cycloalkanes), and cracking, r_2 (r_2 is the rate of conversion to C_6 and lower carbon number alkanes), for two bimetallics (alumina-supported Pt-Re and Pt-Ir) were compared against the performance of Pt-Al_2O_3 catalyst.[112] The *n*-heptane contained 0.5 ppm sulfur, and the reaction rates were determined at low conversion level after 40 hr on stream. The rate (mole per hour per gram catalyst) of dehydrocyclization of *n*-heptane for the Pt-Ir catalyst ($r_1 = 0.05$) was more than twice as high as the rate for the platinum catalyst (0.02), and almost twice as high as the rate for the Pt-Re catalyst (0.03). In contrast, the rate of cracking was higher for the Pt-Ir catalyst. Nevertheless, the selectivity (r_2/r_2) of the Pt-Ir catalyst (0.55) was higher than that of Pt-Re (0.44) and platinum alone (0.42). Excellent correlation of activity and selectivity was obtained with Ir agglomeration in both *n*-heptane/hydrogen and naphtha reforming.[113] The reaction of MCP[112] and cyclohexane[113] over the same catalysts revealed that the Pt-Ir catalyst was less selective for benzene formation than the Pt-Re catalyst or platinum catalyst.

In summary, the model compound-reforming studies by Exxon workers[112,113] have shown that individual hydrocarbon conversion rates are markedly dependent upon the modifier metal of a bimetallic reforming catalyst. In general, the maintenance of selectivity for benzene formation with time-on-stream was better for the Pt-Re catalyst than for the platinum catalyst or Pt-Ir catalyst. The lower benzene selectivity exhibited by Ir additive was attributed to the hydrocracking of the cyclohexane and MCP rings. Pt-Ir catalyst was, however, superior to Pt and Pt-Re catalysts, in activity as well as in selectivity for *n*-heptane dehydrocyclization. The hydrocracking activity of Ir appeared to have been reduced significantly by the addition of sulfur in feed (0.1 to 10 ppm).

As stated above, conversion of *n*-hexane has also often served as a model for studying paraffin reactions. Thus, Parera et al.,[116,126-131] using *n*-hexane and several other reforming reactions including naphtha, showed that the addition of Re and S modified the hydrogenolysis-dehydrogenation capacity of Pt. The role of Re was to decrease the dehydrogenating capacity of Pt, thus reducing coke formation and increasing the formation of lower molecular weight paraffins by hydrogenolysis. The addition of S was influential in decreasing the hydrogenolytic capacity Pt-Re and assisted in increasing the production of benzene and coke. Both geometrical and electronic effects were assigned to influence the changes in selectivity. An electronic effect was also invoked by Boliver et al.[117] to describe the synergism observed for hydrogenolysis-dehydrogenation by adding rhenium to platinum.

Many investigators have used the dehydrogenation of methylcyclohexane (MCH) to toluene to characterize the dehydrogenation activity of the bimetallic reforming catalysts, as well as to determine how poisons influence catalytic activity.[4,110,111,118-121,142] In naphtha reforming, the dehydrogenation of alkylcyclohexanes to aromatics is a major reaction and involves only the hydrogenation-dehydrogenation component of the catalyst. The partial substitution of Pt by Re enhanced the adsorption of toluene to such an extent that it became competitive.[121] This was a result of the preferential adsorption of sulfur on rhenium, keeping the platinum surface free of sulfur; when toluene strongly adsorbed on the platinum surface of a bimetallic cluster.

An extensive study of the effect of Re addition on reforming reactions showed that the rates for production of toluene from MCH and of benzene from MCP and the rate of self-deactivation by MCH (short-term) were unaffected.[118] In contrast, the rate of MCP ring opening decreased, as did the rates for toluene dealkylation and self-deactivation by MCH (long-term). The latter reactions are believed to be structure sensitive while the former are not, suggesting that Re directly affects Pt, as would be expected for Pt-Re clusters formed.

Tin has also received attention as a modifier metal to stabilize the activity of Pt-reforming catalysts. Burch and Garla[93] compared the activity of a Pt monometallic catalyst with and without tin in cyclohexane conversion after extended running with *n*-hexane under the same conditions. This reaction system was chosen so as to probe the metal sites directly. These authors inferred that tin modified both the acidity of the support as well as the Pt electronically with the result that self-poisoning was reduced. In *n*-heptane conversion, Sn did not decrease and at equal loading actually increased coke accumulation vs. Pt monometallic catalysts.[94] It is difficult to draw any conclusion from this study as to how tin stabilizes the platinum catalysts.

MCH dehydrogenation was also used as a probe reaction to study the role of tin and its effect during deactivation on Pt-Sn-Al_2O_3 catalyst.[111] Accelerated deactivation was created by passing MCH over Pt-Sn-Al_2O_3 catalyst after reduction treatment. After deactivation, the catalyst behaved as a physical mixture of the metals rather than a bimetallic system.

The Rutherford backscattering of the deactivated sample in comparison with the reduced catalyst showed certain interesting features (Pt-Sn system has been relatively easy to characterize by RBS as opposed to Pt-Re where it was not possible to distinguish Pt and Re separately because of the limitation on the proximity of their mass numbers). RBS results (Figures 12 and 13) revealed that the Pt/Sn atomic concentration for the near surface layer was larger for deactivated catalyst (0.13) as compared to the reduced catalyst (0.04). It has also been reported that Sn addition decreased Pt particle size in the presence of high platinum loading.[21] In view of the expectation that Pt will concentrate in the surface of Pt-Sn clusters, we might anticipate that Pt and Sn would segregate from one another to an increasingly greater extent as the clusters become smaller and the ratio of surface atoms to the total metal atoms increases.

The very difficult problem of the detailed nature of the carbon deposits formed under reforming is now beginning to be studied. Earlier IR studies[123] indicated that Re may affect the nature of coke in reforming. A recent work[124] utilizing laser Raman spectroscopy showed that the coke on Pt-Re catalysts is much less graphitic than that formed on Pt, Pt-Ir, or Pt-Sn. Neither study examined coke on metal sites specifically.

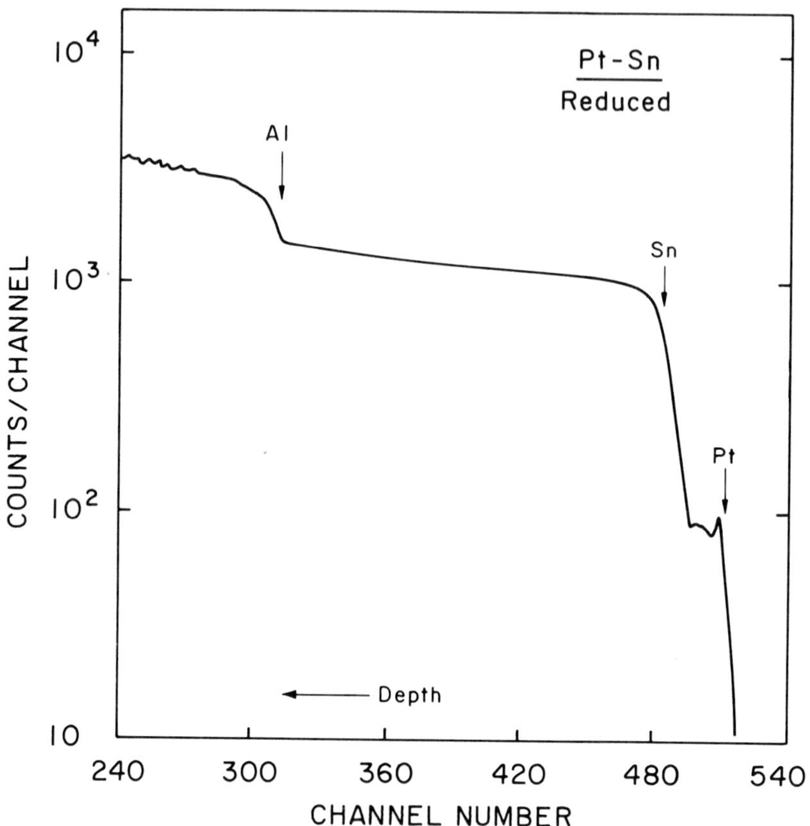

FIGURE 12. RBS spectra of a reduced Pt-Sn-Al_2O_3 catalyst.

Regarding the mechanism of coke formation, several authors have found that MCP and other compounds with five carbon atom rings are very important coke precursors,[131,133] supporting the long-held notion that MCP is an intermediate in coke formation.[34] Bertolacini and Pellet[80] have shown that Re plays the role of somewhat catalyzing the removal of a crucial coke precursor, cyclopentadiene. According to a proposed mechanism,[126] MCP is dehydrogenated to methycyclopentadiene (MCPde) which by dehydrocondensation produces methylindene. The latter can continue dehydrocondensation to a polycyclic compound. High hydrogen pressures are used during the commerical process to keep the formation of indenic structures and the catalyst deactivation very low.

B. Further Remarks on Bimetallic Catalysis

The results of the three bimetallic catalysts provided a good illustration of the tendency of platinum and rhenium, iridium, or tin atoms to form bimetallic clusters, rather than to form monometallic clusters consisting exclusively of atoms of one or the other of the primary and modifier metals.

Recent surface science studies suggested that aromatization activity was most strongly associated with terraces and hydrogenolysis with all sites. Studies of supported Ir catalysts also find little particle-size dependence on the rate of hydrogenolysis of hydrocarbons.[22] Therefore, one can follow the suggestions from the EXAFS and Mössbauer studies as well as from the surface science studies, similar to the one suggested recently,[122] that the Pt-Ir catalyst consists of bimetallic clusters with Pt occupying the sites of lowest coordination, blocking cracking more than aromatization.

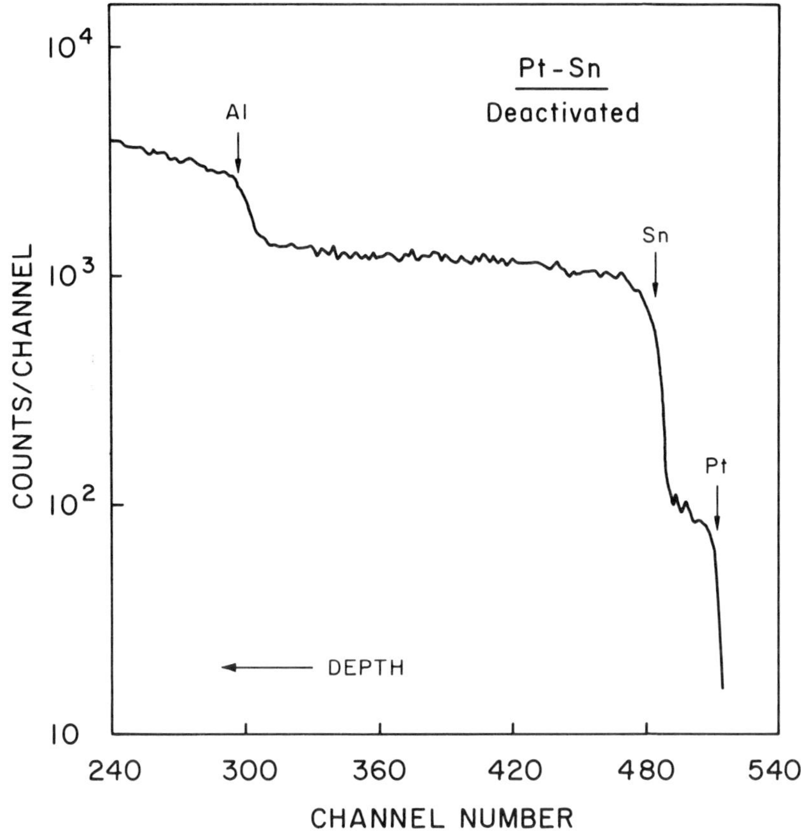

FIGURE 13. RBS spectra of a deactivated Pt-Sn-Al_2O_3 catalyst.

On the basis of the characterization and reaction findings taken together, it appears that the state of Pt-Re-Al_2O_3 catalysts is sensitive to the conditions employed in their preparation. Under some conditions, there appears to be strong evidence of significant interaction between the platinum and rhenium to form bimetallic clusters. With other conditions, the interaction between the two elements seems to be less pronounced. The modern catalysts seek to combine the two metals in bimetallic clusters to obtain superior stability. The chemisorbed sulfur atoms are present on surface rhenium atoms which in turn remain bonded to platinum atoms in the surface region. The combined effect of sulfur and rhenium divides the active metal into ensembles too small to permit coke formation. It must also be pointed out that the sensitivity of the catalysts to conditions of preparation and activation introduces considerable uncertainty in projections about the nature of the catalyst under actual reforming conditions.[3]

In spite of the large number of investigations on Pt-Sn-Al_2O_3 catalysts, both characterization and reaction studies, the exact role of tin in stabilizing the catalyst activity of platinum is still not clear. However, the Mössbauer and RBS studies have proven useful in helping to think about stabilization mechanism specific to Pt-Sn catalysts. It appears that both geometrical and electronic effects influence the changes in activity and selectivity. However, to reconcile with an electronic effect, we need more evidence, but specific to the point, which is not available at this stage.

C. Extended Naphtha Reforming

Results of some extended naphtha reforming runs will illustrate further the differences between platinum-iridium and platinum-rhenium on alumina catalysts. A comparison of

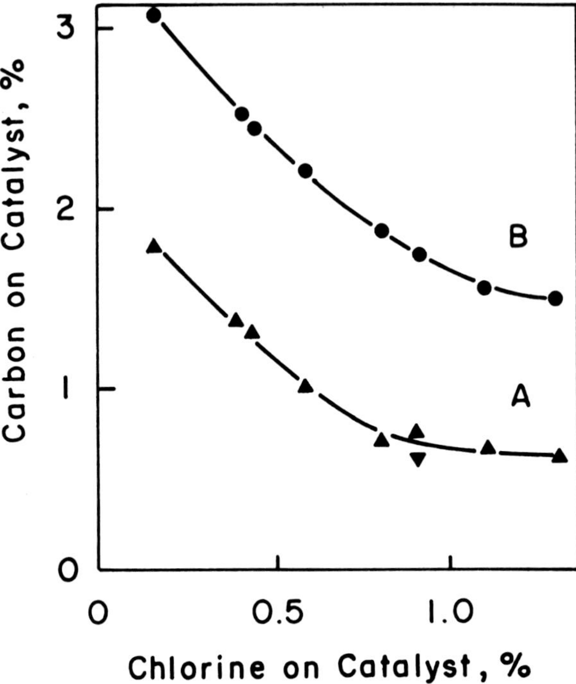

FIGURE 14. Carbon percentage on a commercial Pt-Re-Al_2O_3 catalyst as a function of the chlorine concentration: A, unsulfided; B, sulfided. (From Parera, J. M., Verderone, R. J., Pieck, C. L., and Traffano, E. M., *Appl. Catal.*, 7, 43, 1983. With permission.)

these catalysts with platinum-tin on alumina catalyst is also given. However, this review deals mostly with the two bimetallic catalysts, Pt-Ir and Pt-Re, as some pilot plant data are available. Most of these studies have been carried out with sulfurized and chlorided catalysts, in accordance with commercial practice. The discussion does not include the effects of alkali metal doping. Again, of the commercial reforming studies published so far, those by the Exxon researchers[3,10,112] are particularly comprehensive. We will begin with catalyst activity and maintenance of activity during the initial period of reforming.

Chlorided bimetallic catalysts have an initial period with great gas production and poor liquid yield. During this period, the main reaction is hydrogenolysis, a highly exothermic reaction. This produces a runaway condition, known as runaway hydrocracking, which in turn brings reactor instability. In industrial practice, this hydrogenolytic hyperactivity is passivated by sulfurizing the catalysts. It does decrease the hydrogenolytic capacity of Pt-Re but significantly increases the production of coke. Figure 14 shows the amount of coke deposited on sulfurized and unsulfurized commercial Pt-Re-Al_2O_3 catalysts as a function of the chlorine concentration after deactivation test.[130] Accelerated deactivation tests were carried out using a commercial hydrotreated naphtha of the boiling range 338 to 420 K. It can be observed that increase in the concentration of chlorine results in decrease of the coke deposition, and a net increase in carbon deposition when the catalyst was sulfurized.

Pt-Sn catalysts accumulate as much and perhaps more coke than the monometallic even though activity is maintained.[94] Pt-Ir is more active than Pt for aromatization, but also, unfortunately for cracking, and usually gives higher yields of methane and ethane than that of Pt-Re catalysts. The yield of C_5+ reformate obtained with a Pt-Re catalyst is usually higher (typically 1 vol% for wide boiling-range naphthas), however, under certain conditions with light naphtha fractions, this result is reversed.[10]

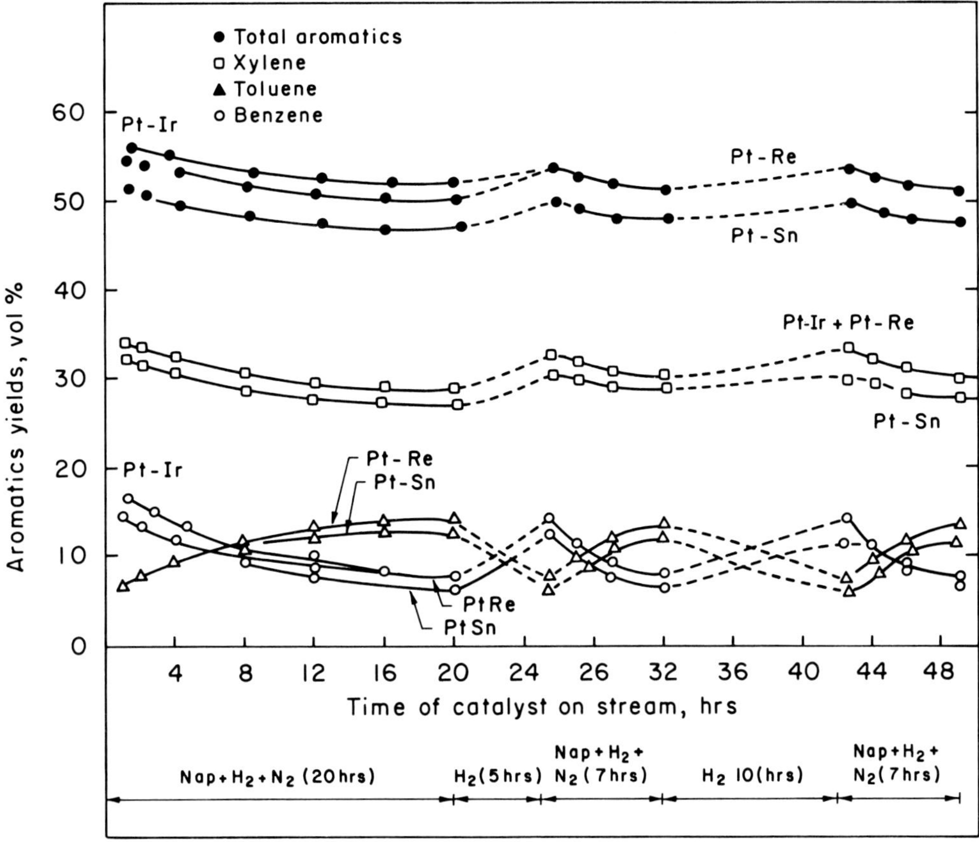

FIGURE 15. Comparison of aromatic yields vs. time of catalyst on stream obtained with alumina supported Pt-Re, Pt-Ir, and Pt-Sn catalysts in the reforming of a 320 to 430 K boiling-range naphtha. Catalysts contained 0.3 wt% each metal.

The yields of specific aromatic hydrocarbons in the reformate are also of interest. Reforming of light naphtha (320 to 430 K) at 763 K was carried out with Pt-Re, Pt-Ir, and Pt-Sn catalysts for 20 hr. A naphtha to hydrogen to nitrogen mole ratio of 1:7:7 was used for the reforming reactions; nitrogen was used in the feed to accelerate deactivation. At the end of the 20-hr period, the reforming was discontinued and the catalyst was regenerated by flowing a stream of pure H_2 for 5 hr. The cycle was repeated once again.

Typical laboratory data for the aromatic yield as a function of time are shown in Figure 15. Presence of nitrogen lowers hydrogen partial pressure which perhaps favors the dehydrogenation of paraffins to coke precursors and subsequent coverage of the catalyst surface by carbonaceous residues, and decreases the reaction rate. During the regeneration cycle (21 to 25 hr of operation) the coke is hydrogenated to produce gaseous hydrocarbons increasing the catalyst activity to nearly the original value. The second and third reformation periods which are 7 hr duration each showed similar deactivation. While first regeneration was of 5 hr duration, the second-time regeneration was continued for 10 hr. From a general comparison of aromatic yields at the end of the regeneration periods, it appears that 5 hr is sufficient to remove the coke. In comparing the aromatic yields for Pt-Ir and Pt-Re catalysts with Pt-Sn catalyst, we see that the yields of xylenes, toluene, benzene, and total aromatic hydrocarbons in the liquid reformate are consistently higher for the Pt-Ir and Pt-Re catalysts. Between the Pt-Re and Pt-Ir catalysts, the yields are comparable for the extended period.

With regard to commercial regeneration, Pt-Re can take several regeneration cycles as

Table 8
COMPARISON OF IRIDIUM-CONTAINING TRIMETALLIC WITH IRIDIUM-BIMETALLIC PHYSICAL MIXTURE[137]

	Pt-Re-Ir trimetallic	Pt-Re bimetallic + Ir monometallic
Cyclohexane conversion (%)	79.1	86.9
n-Hexane conversion (%)	51.2	52.9
Effluent methane content (%)	2.8	15.9
C_5+ liquid yield in reforming (vol%)	74.8	71.3
Below C_5+ reforming yield (wt%)	18.0	22.0

Note: Catalysts: 0.3 wt% each metal, CO_2 pretreatment, H_2 reduced at 727 K, gamma alumina support. Cyclohexane dehydrogenation conditions: 0.5 g catalyst, WHSV = 3, H_2/HC = 7, 570 K, 10 min. *n*-Hexane dehydrocyclization; 0.5 g catalyst; same catalyst as cyclohexane dehydrogenation after H_2 flush, WHSV = 3, H_2/HC = 7, 740 K, 60 min. Reforming: naphtha — 71.8% paraffins, 17.4% naphthenes, 10.8% aromatics, 2 ppm S; 98 octane product; yield measured after 30 days on oil.

compared to Pt-Ir. During the coke burn of Pt-Ir catalysts in an oxygen-containing atmosphere near 770 K, Ir is more susceptible toward oxidative agglomeration than the platinum component.[87,112] Ir oxidizes to IrO_2 in the neighborhood of 600 K. Large particles of the oxides form from small original metal particles during high-temperature burn off. Direct reduction gives large Ir particles. To minimize the oxidative agglomeration, regeneration procedures employing repeated treatments with oxygen/chlorine burn mixture have been developed to obtain high metal dispersion.[134-136] In a recent paper, McVicker and Ziemiak[113] reported the influence of Ir agglomeration on the naphtha-reforming behavior, and obtained excellent correlation of activity and selectivity with Ir agglomerates. However, these regeneration and redispersion practices have been a constant concern to the industry.

In comparing the product distribution, the data for the reactions of the selected hydrocarbons considered earlier indicate that Pt-Re catalysts are particularly effective for the selective conversion of cycloalkanes into aromatic hydrocarbons, whereas Pt-Ir catalysts are highly active for the conversion of noncyclic alkanes into aromatics. Therefore, an improved process can result when Pt-Re and Pt-Ir are used in combination.

The behavior of mixed beds was indeed revealing for Pt-Re-Ir system and is worth discussing here. Table 8 compares the performance in a series of reactions of two alumina-supported Pt-Re-Ir trimetallic catalysts, one where the Ir is alloyed and the other where it is incorporated as separately supported Ir.[137] Its high activity is especially evident for the mixture, but its tendency for hydrogenolysis is still striking.[122]

Exxon workers[112,138] also performed a combined catalyst operation in which a platinum-rhenium catalyst was used in the front of the system, followed with a platinum-iridium catalyst. Reforming data on a 335 to 460 K boiling-range naphtha are given in Table 9.[112] In the combined catalyst system, the platinum-iridium catalyst constitutes 65% of the total catalyst charge. The data show that the Pt-Ir catalyst is approximately twice as active as the Pt-Re catalyst. In comparing the activity for the Pt-Ir and Pt-Re catalysts with the combined catalyst, one can see that the activity of the combined catalyst system is intermediate between the activities of the individual Pt-Ir and Pt-Re catalysts. The yields of C_5+ reformate are equivalent for the Pt-Re catalyst and the combined catalyst system and about 1.0 to 1.5 vol% higher than for the Pt-Ir catalyst. Also, methane and ethane yields for the combined system were reported to be higher than those for the Pt-Re catalyst but lower than those for the Pt-Ir catalyst.

In practice, a reformer consists of a numer of reactors in series. Since the aromatization of cycloalkanes occurs primarily in the first region (upstream) whereas aromatization of noncyclic alkanes occurs predominantly in the second stage (downstream), a combination

Table 9
COMPARISON OF ACTIVITIES AND C_5+ REFORMATE YIELDS (VOL%) OF ALUMINA-SUPPORTED Pt-Ir AND Pt-Re CATALYSTS WITH THAT OF A COMBINED CATALYST SYSTEM

	Pt-Ir		Pt-Re		Combined catalyst	
Hr on stream	Activity	C_5+	Activity	C_5+	Activity	C_5+
200	160	69	90	70	120	70
500	170	69	80	70	150	71
1000	160	68	70	70	110	70
1500	150	68	65	70	100	70

Note: The combined catalyst constitutes 65 wt% of the total catalyst charged. Data are for the reforming of a 335 to 460 K boiling-range naphtha at 770 K and 28.2 atm to produce 98 research octane number reformate.

From Carter, J. L., McVicker, G. B., Weissman, W., Kmak, W. S., and Sinfelt, J. H., *Appl. Catal.*, 3, 327, 1982. With permission.

of Pt-Ir and Pt-Re can be used in the commercial reformers. The Pt-Re catalyst is more selective for the reforming reactions occurring in the upstream sections of the series of reactors, and the Pt-Ir catalyst more selective for those occurring in the downstream sections. A number of commercial reforming operations now employ combined catalyst systems of the type described here.

Chevron researchers[82,97] have also been active in developing a zeolite catalyst for naphtha reforming. Their recently announced catalyst,[97] which contains highly dispersed platinum clusters in barium-exchanged potassium Zeolite L, is active and selective for the aromatization of paraffins, especially heptanes and hexanes, as well as for alkylcyclopentane aromatization. Unlike the bifunctional reforming catalyst, the Pt/BaKL catalyst, is nonacidic and catalyzes aromatization using only the catalyst properties of the platinum clusters. The catalyst was reported to be extremely sensitive to poisoning by sulfur.

VI. REACTION ENGINEERING

The reactor operation is now considered more specifically in terms of the reforming catalytic chemistry. Although the new reformers operate with bimetallic catalysts, the kinetic information and deactivation parameters are scarce due to proprietary reasons. The industrial models have been developed which predict reforming performance very well, but only simple versions of these models, especially with the kinetic parameters related to Pt-Al_2O_3 catalysts, have been published. More realistic kinetic information and deactivation data over an extended period with bimetallic catalysts are required in order to incorporate them in a quantitative modeling of a commercial reforming system. With limited information, we, however, proceed to illustrate the approach of reaction engineering here by developing a simple procedure to extract kinetic parameters for reforming reactions from a detailed set of kinetics equations which lead ultimately to the quantitative modeling of a three-reactor commercial reforming system.

We begin here with an overview of some recently published kinetic and deactivation rate data obtained from model compounds using bimetallic catalysts. The reactor modeling then follows.

Table 10
FIRST-ORDER RATE CONSTANTS FOR NAPHTHA-REFORMING REACTIONS[140]

Dehydrocyclization k_1	$k \times 10^2$, h^{-1}	Hydroisomerization of aromatics k_4	$k \times 10^2$, h^{-1}
$P_{10} \rightarrow N_{10}$	2.54		
$P_9 \rightarrow N_9$	1.81	$A_{10} \rightarrow P_{10}$	0.16
$P_8 \rightarrow N_8$	1.33	$A_9 \rightarrow P_9$	0.16
$P_7 \rightarrow N_7$	0.58	$A_8 \rightarrow P_8$	0.16
$P_6 \rightarrow N_6$	0	$A_7 \rightarrow P_7$	0.16
Dehydrogenation k_2		Hydrocracking of naphthenes k_5	
$N_{10} \rightarrow A_{10}$	24.50	$N_{10} \rightarrow N_9$	1.34
$N_9 \rightarrow A_9$	24.50	$N_{10} \rightarrow N_8$	1.34
$N_8 \rightarrow A_8$	21.50	$N_{10} \rightarrow N_7$	0.80
$N_7 \rightarrow A_7$	9.03	$N_9 \rightarrow N_8$	1.27
$N_6 \rightarrow A_6$	4.02	$N_9 \rightarrow N_7$	1.27
Hydrocracking of paraffins k_3		$N_8 \rightarrow N_7$	0.09
$P_{10} \rightarrow P_9 + P_1$	0.49		
$P_{10} \rightarrow P_8 + P_2$	0.63	Hydrogenation k_6	
$P_{10} \rightarrow P_7 + P_3$	1.09	$A_6 \rightarrow N_6$	0.45
$P_{10} \rightarrow P_6 + P_4$	0.89		
$P_{10} \rightarrow 2P_5$	1.24	Hydrocracking of aromatics k_7	
$P_9 \rightarrow P_8 + P_1$	0.30	$A_{10} \rightarrow A_9$	0.06
$P_9 \rightarrow P_7 + P_2$	0.39	$A_{10} \rightarrow A_8$	0.06
$P_9 \rightarrow P_6 + P_3$	0.68	$A_{10} \rightarrow A_7$	0.00
$P_9 \rightarrow P_5 + P_4$	0.55	$A_9 \rightarrow A_8$	0.05
$P_8 \rightarrow P_7 + P_1$	0.19	$A_9 \rightarrow A_7$	0.05
$P_8 \rightarrow P_6 + P_2$	0.25	$A_8 \rightarrow A_7$	0.01
$P_8 \rightarrow P_5 + P_3$	0.43		
$P_8 \rightarrow 2P_4$	0.35	Hydroisomerization of naphthenes k_8	
$P_7 \rightarrow P_6 + P_1$	0.14		
$P_7 \rightarrow P_5 + P_2$	0.18	$N_{10} \rightarrow P_{10}$	0.54
$P_7 \rightarrow P_4 + P_3$	0.32	$N_9 \rightarrow P_9$	0.54
$P_6 \rightarrow P_5 + P_1$	0.14	$N_8 \rightarrow P_8$	0.47
$P_6 \rightarrow P_4 + P_2$	0.18	$N_7 \rightarrow P_7$	0.20
$P_6 \rightarrow 2P_3$	0.27	$N_6 \rightarrow P_6$	1.48

Note: Typical reaction conditions: 496°C, hydrogen to naphtha molar ratio = 5:1, 6.8 to 30 atm total pressure, and 0.30 wt% Pt on alumina catalyst. P, N, and A denote paraffin, naphthene, and aromatic, respectively, and the subscript is the carbon number.

A. Rate Expressions

The kinetics of model-compound reforming studies with bimetallic catalysts have been carried out recently.[112,119,121] The deactivation-kinetic analysis has also been slow to develop. It is only very recently that some deactivation-rate data have been obtained.[118,120,139] Again, $Pt\text{-}Re\text{-}Al_2O_3$ catalyst has received the most attention. Kinetics of reforming reactions on $Pt\text{-}Al_2O_3$ have already been the subject of several excellent reviews.[18,140] Table 10 summarizes first-order rate constants for 53 individual reactions characteristic of naphtha reforming. The rate constant for each reaction is defined by

$$\frac{\partial N_i}{\partial(1/SV)} = -k_i N_i \tag{1}$$

where N_i denotes moles of reacting component i and SV indicates weight liquid hourly space velocity. Recently, McVicker et al.[141] obtained kinetic data with the reforming of model compounds (cyclohexane, methylcyclopentane, and *n*-heptane) over $Ir\text{-}Al_2O_3$ catalyst.

Table 11
PARAMETER ESTIMATES FOR RATE EXPRESSION (2) ON Pt-Re-Al_2O_3

Kinetic parameter	Value
$A_{o,\,MCH\text{-}Tol}$, kmol/(kg of cat. hr atm)	2.92×10^{13}
$E_{MCH\text{-}Tol}$, kJ/mol	181.4
K_{MCH}, atm^{-1}	1.75
K_{Tol}, atm^{-1}	18.1
K_{np7}, atm^{-1}	1.86

From Van Trimpont, P. A., Marin, G. B., and Froment, G. F., *Ind. Eng. Chem. Fundam.*, 25, 544, 1986. With permission.

The two most recent studies[119,121] described the detailed kinetics of dehydrogenation of methylcyclohexane in the presence of added hydrogen over Pt-Re-Al_2O_3 catalyst under conditions of no mass-transfer influence. Their experimental results were analyzed on the basis of Langmuir-Hinshelwood kinetics using statistical data interpretation. While Srivastava and co-workers[119] performed their study at atmospheric pressure, Van Trimpont et al.[121] carried their investigation under quasi-industrial conditions. In both cases, a single-site mechanism was favored. The zero reaction order, with respect to hydrogen, and the negative order, with respect to toluene, can be understood from the Hougen-Watson rate expression. This requires the concentrations of adsorbed H_2 and dehydrogenated reaction intermediates to be small when compared with the concentrations of adsorbed MCH and toluene. In the experiments of Van Trimpont et al.,[121] an estimation of the adsorption coefficient of the *n*-heptane added to the feed was possible.

The data were most satisfactorily correlated by the rate expression[121]

$$r_{MCH\rightarrow Tol} = \frac{k_{MCH\rightarrow Tol}(p_{MCH} - p_{Tol}\, p_H^3/K_{MCH\rightarrow Tol})}{1 + K_{MCH}\, p_{MCH} + K_{Tol}\, p_{Tol} + K_{np7}\, p_{np7}} \tag{2}$$

The parameter estimates are shown in Table 11.

Srivastava and co-workers[111,120] also obtained the data for kinetics of deactivation by coking of Pt-Re-Al_2O_3 and Pt-Sn-Al_2O_3 catalysts for the dehydrogenation of methylcyclohexane. The kinetic data for dehydrogenation and deactivation were analyzed on the basis of Langmuir-Hinshelwood models to develop statistically best rate expressions. The deactivation analysis was accomplished by using the procedure described in Chapter 1. In both cases, deactivation occurred in parallel with the main reaction (MCH → Tol) where MCH was absorbed in different ways in the main and the deactivation reactions. The model which described the deactivation rate, $(-da/dt)$, for Pt-Re-Al_2O_3 most satisfactorily was

$$r_{deact} = \frac{k_d\, K_{MCH^*}^2\, p_{MCH}^2}{(1 + K_{Tol}\, p_{Tol} + K_{MCH^*}\, p_{MCH})^2} \cdot a^2 \tag{3}$$

where ‘a’ was the catalyst activity.[120] This equation was governed by the reaction of two adjacent adsorbed methylcyclohexane molecules, resulting in the coke precursor. For Pt-Sn-Al_2O_3 catalyst, the deactivation rate was correlated by

$$r_{deact} = \frac{k_d\, K_{MCH^*}\, p_M}{(1 + K_{H_2}\, p_{H_2} + K_{MCH^*}\, p_{MCH})} \cdot a \tag{4}$$

Table 12
PARAMETER ESTIMATION FOR DEACTIVATION RATE EXPRESSIONS (3) AND (4)

	Value	
Kinetic parameter	**Equation (3)**	**Equation (4)**
$(k_d)_o$, hr^{-1}	5.64×10^{11}	1.1×10^{3}
E_{deact}, kJ/mol	146.6	53.2
K_{MCH^*}, atm^{-1}	0.014	7.8×10^{3}
K_{H_2}, atm^{-1}	—	12.7
K_{Tol}, atm^{-1}	1.59×10^{-4}	—

In this case, a significant estimation of the adsorption coefficient of the hydrogen added was possible. The parameter estimates for both the systems are summarized in Table 12. In contrast to Pt-Re deactivation equation, the above expression for Pt-Sn shows that an increase in hydrogen partial pressure has a retarding effect on the formation of coke, thereby increasing the activity maintenance, whereas the addition of toluene does not affect the deactivation process.

B. Modeling

Naphtha is a complex mixture of hydrocarbons with at least 285 components identified in certain naphthas.[143] Moreover, one naphtha may differ significantly from another in its composition. With such a large number of components, the task for accounting for all of them directly in any kinetic model becomes exceedingly difficult. Hence, it is imperative to group these chemical components into a smaller set of kinetic lumps.

A large number of kinetic models for catalytic reforming is available in published literature. These include models by Smith,[144] Krane et al.,[140] Henningsen and Boundgaard,[145] and the Mobil model by Ramage et al.[143] The Mobil model is the latest in line, and is reported to have achieved excellent results. It is also the most complex. In this model, 'reforming kinetics through a minimal number of lumped species and a minimal number of reaction pathways between these lumps such that the resulting kinetic parameters are independent of feedstocks composition' is envisaged. The data obtained with over 300 material balances, five commercial catalysts, different temperatures, pressures, hydrogen to hydrocarbon ratio, liquid hourly space velocities, and charge stocks were used to estimate five kinetic constants at each step, thereby reducing the errors involved.

On the other hand, Smith's model has the smallest number of lumps viz. naphthenes, paraffins, and aromatics. Its simplicity turns out to be extremely advantageous in the wake of paucity of data. We begin with a typical example where Smith's model has been used to predict plant performance.[146]

In Smith's model the complex naphtha mixture is idealized so that each of the hydrocarbon classes, naphthenes (cycloparaffins), paraffins, and aromatics, is represented by a single compound having average properties of that class. It is assumed that each of these three compounds has the same number of carbon atoms, n_c. This assumption is justified on the ground that the precursor and the corresponding product have the same number of carbon atoms, e.g.,

$$\underset{(n_c = 7)}{\text{Heptane}} \rightleftharpoons \underset{(n_c = 7)}{\text{Methylcyclohexane}} \rightleftharpoons \underset{(n_c = 7)}{\text{Toluene}} \tag{5}$$

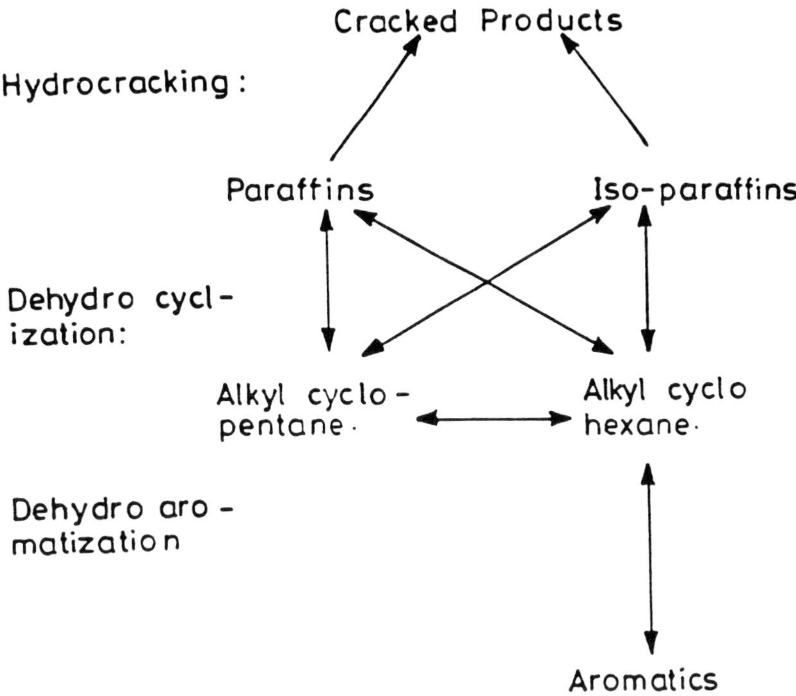

FIGURE 16. Simplified reaction network for catalytic reforming.

$$\text{Octane} \rightleftharpoons \text{di-Methylcyclohexane} \rightleftharpoons \text{Xylene} \tag{6}$$

Some disproportionation may be caused by the hydrocracking of paraffins and naphthenes but both are much slower when compared with the other reactions. Of the many reactions described in Section II.A (Figure 1), the following major ones are taken into consideration:

1. Dehydrogenation of naphthenes to aromatics
2. Dehydrogenation of paraffins to naphthenes
3. Hydrocracking of naphthenes to lower hydrocarbons (C_1 to C_5)
4. Hydrocracking of paraffins to lower hydrocarbons (C_1 to C_5)

A simple network showing the above reactions is given in Figure 16.[147]

Ideally, the model should have provisions to individually account for the activity decline of Pt and Al_2O_3. But since the plant data available were from a fresh catalyst and were not over an extended period, both the declines were lumped together as a single parameter varying between 1 and 0 (1 for a fresh catalyst). It has been assumed that a cycle lasts for 6 months, at the end of which the catalyst is regenerated. At the end of each cycle the activity declines to 93% of the value at the start of the cycle and after regeneration, regains 98% of the start of the cycle value.

Though the reactions are heterogeneously catalyzed, to further simplify the calculations, they are assumed to follow homogeneous kinetics which seems to be a normal feature of all models. Thus,

Reaction I:

$$\text{Naphthene (1 mol)} \underset{k_r}{\overset{k_f}{\rightleftharpoons}} \text{Aromatics (1 mol)} + 3H_2 \tag{7}$$

$$-\frac{dN_N}{dV_R} = \frac{k_f}{K_I}(K_I\, p_N - p_A\, p_{H_2}^3) \tag{8}$$

Reaction II:

$$\text{Naphthene (1 mol)} + H_2 \rightleftharpoons \text{Paraffins (1 mol)} \tag{9}$$

$$-\frac{dN_N}{dV_R} = \frac{k_f}{K_{II}}(K_{II}\, p_N\, p_{H_2} - p_P) \tag{10}$$

Reaction III:

$$\text{Naphthenes} \rightarrow \text{Lighter ends} \tag{11}$$

In general, the reforming plant reveals that methane, ethane, propane, butane, and pentane are produced in approximately equal molar proportions in the hydrocracking reactions. This is generally termed as 'invarient partitioning of hydrocracking products'. Stoichiometric considerations, coupled with this fact, help in writing down the chemical reaction as

$$C_n\, H_{2n+2} + \left(\frac{n-3}{3}\right) H_2 \rightarrow \frac{n}{15}\,[C_1 + C_2 + C_3 + C_4 + C_5] \tag{12}$$

Hydrocracking reaction rates are first order with respect to the precursor. This gives the raction rate as

$$-\frac{dN_N}{dV_R} = \frac{K_c}{p_T}\, p_N \tag{13}$$

Reaction IV:

$$\text{Paraffins} \rightarrow \text{Lighter ends} \tag{14a}$$

Similar treatment as meted for naphthene hydrocracking is carried out for paraffin hydrocracking also. Thus,

$$-\frac{dN_p}{dV_R} = \frac{K_c}{p_T}\, p_p \tag{14b}$$

where

- N_N, N_p = the number of moles of naphthenes and paraffins, respectively
- p_N, p_A, p_p, p_{H_2} = partial pressure of naphthenes, aromatics, paraffins, and hydrogen, respectively
- K_I, K_{II} = equilibrium constants
- V_R = volume per time

Table 13
PLANT DATA: FEEDSTOCK AND RESULTS OF COMMERCIAL RUN

Components	Compositions (mol%)
Feedstock	
Paraffins	50
Naphthenes	35
Aromatics	15
Reformate analysis	
Nonaromatics	33.3
Benzene	0.8
Toluene	13.0
Ethylbenzene	7.2
p- and *m*-Xylene	28.9
o-Xylene	9.8
Cumene	0.1
C9 + HC	6.9
Recycle gas flow	
C1	4.7
C2	4.2
C3	0.9
C4	2.1
C5	0.8
C6	0.2
H2	87.0
Off-gas	
C1	2.6
C2	11.2
C3	34.9
C4	41.2
H2	10.0

- K_c = hydrocracking reaction rate constant
- p_T = total pressure

In order to include temperature effects, an energy balance must now be added. As far as the heats of reaction are concerned, a whole range of values has been reported. However, heats of reaction are nearly independent of the component molecular weights over the range normally encountered in reforming when they are based on the number of moles of hydrogen entering into the reaction. For the present case, the values reported by Smith[144] have been used. Smith also suggests that the activation energies for paraffin hydrocracking and naphthene hydrocracking remain the the same. For adiabatic operation the enthalpy balance for the system is given by

$$dN_{N\to p}\,\Delta H_{N\to p} + dN_{N\to A}\,\Delta H_{N\to A} + dN_{p\to LHC}\,\Delta H_{p\to LHC} + dN_{N\to LHC}\,\Delta H_{N\to LHC} = N_T(MW)_R\,C_p\,dT \tag{15}$$

where $(MW)_R$ is the molecular weight of the mixture and N_T is the total molar flow rate.

We now consider the modeling of the three-reactor reforming system shown in Figure 2. Only limited data are available regarding the composition of the naphthas in terms of amounts of paraffins, naphthenes, and aromatics. Average carbon number and average molecular weight of naphtha are calculated from available data. Tables 13 and 14 give plant data which include the feedstock (average carbon number, 7.85) and reformate compositions, and assigned operating parameters. The required equations and data are as follows.

Table 14
PLANT DATA: SYSTEM OPERATING PARAMETERS

Operating parameters	
H2/HC, molar flow rate	7.97
Flow rates	
Naphtha, MT/hr	14.09
C_5+ reformate, MT/hr	12.51
Recycle gas, kg mol/hr	1153
Off-gas, kg mol/hr	9.7
Reactor pressure and inlet temperature	
R-1	40 atm and 766 K
R-2	40 atm and 770 K
R-3	490 atm and 771 K
Catalyst load (kg)	
Reactor, R-1	658
R-2	883
R-3	6020

Reaction I:

$$K_I = \exp(46.15 - 50784/RT) \text{ atm}^3 \tag{16}$$

$$(k_f)_I = \exp(23.21 - 32{,}881/RT) \cdot [\text{act}] \tag{17}$$

$$\text{kg mol/hr} \cdot \text{kg} \cdot \text{cat} \cdot \text{atm}$$

where [act] = catalyst's activity.
Rate equation: Equation 8
Energy balance:

$$\Delta H_I = +50{,}820 \text{ cal/gmol naphthene consumed}$$

$$(dT/dV_R)_I = (dN_N/dV_R)_I \frac{50{,}820}{N_T\, C_p(MW)} \tag{18}$$

Reaction II:

$$K_{II} = \exp(8823/RT - 7.12) \text{ atm}^{-1} \tag{19}$$

$$(k_f)_{II} = \exp(35.98 - 63{,}613/RT) [\text{act}] \text{ kg mol/hr} \cdot \text{kg} \cdot \text{cat} \cdot \text{atm} \tag{20}$$

Rate equation: Equation 10
Energy balance:

$$\Delta H_{II} = -10{,}553 \text{ cal/gmol } H_2 \text{ consumed}$$

$$(dT/dV_R)_{II} = (dN_N/dV_R)_{II} \left[\frac{-10{,}553}{N_T\, C_p(MW)_R}\right] \tag{21}$$

Reaction III:

$$(k_f)_{III} = \exp(42.97 - 66{,}750/RT) [\text{act}] \text{ kg mol/hr kg cat} \tag{22}$$

Table 15
COMPARISON OF CALCULATED VALUES WITH PLANT DATA

	Reactor					
	1		2		3	
	Calc	Plant	Calc	Plant	Calc	Plant
Temp (K)	689	689	725	745	757	757

	Overall Results	
	Calc	Plant
Aromatics leaving reactor, wt%	67.1	66.7
Hydrogen (kg mol)	1093	1001
Reformate research octane number (clear)	95.8	95.5

Rate equation: Equation 13
Energy balance:

$$\Delta H_{III} = -12{,}386 \text{ cal/gmol of } H_2 \text{ consumed}$$

$$(dT/dV_R)_{III} = (dN_N/dV_R)_{III} \times n_c/3 \times \frac{-12{,}386}{N_T\, C_p(MW)_R} \tag{23}$$

Reaction IV:

$$(k_f)_{IV} = \exp(42.97 - 66{,}750/RT) \text{ [act] kg mol/hr kg cat} \tag{24}$$

Rate Equation: Equation 15
Energy balance:

$$\Delta H_{IV} = -13{,}496 \text{ cal/gmol}$$

$$(dT/dV_R)_{IV} = (dN/dV_R)_{IV}\, [(n_c - 3)/3]\, \frac{-13{,}496}{N_T\, C_p(MW)_R} \tag{25}$$

The component mass balances and the energy balance are solved by progressively stepping down the reactor. The overall rates of change of naphthenes, paraffins, and aromatics as well as the overall rate of change of temperature along the reactor length are calculated by combining and integrating the contributions of the individual reactions.

Table 15 gives a comparison between calculated and observed values of temperature and product composition. The calculated first and third reactor exit temperature matches very well with the corresponding plant value, but the simulation predicts a lower temperature for the second reactor. While the aromatics composition also matches very well, the model foretells higher hydrogen content. The model, simple as it is, predicts the plant performance reasonably well.

The temperature and concentrations at different points in the reactor are shown in Figure 17. The temperature decreases in an exponential manner and then flattens out in all three

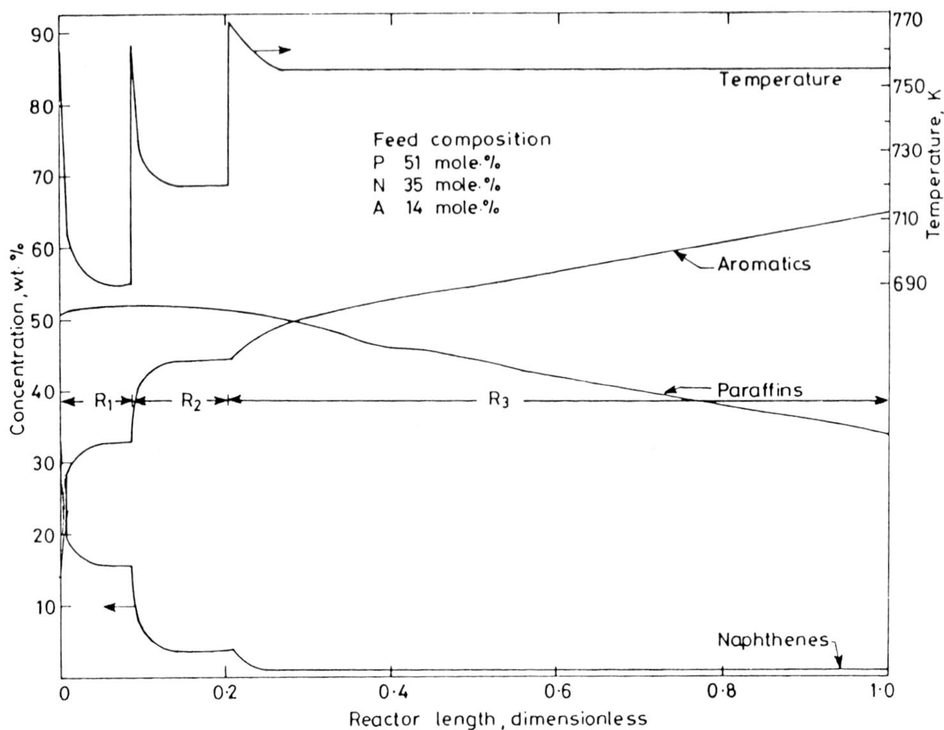

FIGURE 17. Concentration and temperature profiles in the reformers.

reactors. The large temperature drop in the first reactor is indicative of the naphthenes present in the feed. It appears that in the first and second reactors, the aromatics increase exponentially at the cost of the naphthenes. There is an insignificant increase in paraffin content in the initial part of the first reactor and for the rest of the first and second reactors this seems to remain constant. The increase in concentration of aromatics in the third reactor appears to be solely due to the disappearance of paraffins.

Regarding the nexus between temperature and the conversions, the whole reaction set as such is exothermic. The temperature drop in the third reactor being not steep, unlike the first and second, is explained by the fact that some hydrocracking of paraffins, which is exothermic, also accompanies the conversion to aromatics. The large temperature decrease in the first two reactors almost quenches all reaction, and, therefore, it is undesirable to have additional catalyst in the first two reactors since excessive reheating of the gas stream is required. The flat portion of curves for both temperature and composition in the first two reactors is indicative of this phenomenon where part of the catalyst was not effectively used.

VII. STATUS AND OUTLOOK

Catalytic reforming is a firmly established reforming process that will continue to be used. Currently, it appears that the need for expansion of reformer capacity will be limited for at least the next several years, mainly because of a combination of the economic pressures of high gasoline costs and automobile manufacturers' improvements in automobile mileage. The bimetallic catalysts have exhibited outstanding performance in refineries, but many questions about bimetallic catalysis remain unanswered. Some trends in current research are given below.

In recent years, research efforts in this area have shifted from studies of chemical reaction kinetics and mechanisms to investigations of specific surface reactions. EXAFS and Mössbauer will continue to be useful for the elucidation of structural and electronic properties. It would, however, be useful to place more emphasis on characterization in very different types of chemical environments. These studies, in parallel with model studies using well-defined catalysts having one or a few species present, will reduce ambiguities.

The outlook for applications of bimetallic clusters in reaction systems other than reforming is appealing in view of the many possible combinations of metallic elements. The extension to tri- or polymetallic catalysts offers the design of improved catalyst formulations and holds the promise to formulate new catalysts for carrying out specific chemical functions.

Another change that might occur in catalytic reforming is the development of catalysts containing platinum impregnated in alkaline earth zeolites, particularly, Zeolite L. Such catalysts are especially selective for aromatization of *n*-paraffins. Although the catalyst deactivates rapidly in the presence of sulfur, its stability for reforming of thoroughly desulfurized light naphtha has been commercially demonstrated.

REFERENCES

1. **Edgar, M. D.,** in *Applied Industrial Catalysis,* Vol. 1, Leach, B. E., Ed., Academic Press, New York, 1983, chap. 5.
2. **Haensel, V.,** *Am. Chem. Soc. Symp. Ser.,* 222, 144, 1983.
3. **Sinfelt, J. H.,** in *Catalysis Science and Technology,* Anderson, J. R. and Boudart, M., Eds., Springer-Verlag, New York, 1981, chap. 5; and references cited therein.
4. **Jothimurugesan, K., Nayak, A. K., Mehta, G. K., Rai, K. N., Bhatia, S., and Srivastava, R. D.,** *AIChE J.,* 31, 1997, 1985.
5. **Kelley, M. J., Freed, R. L., and Swartzfager, D. G.,** *J. Catal.,* 78, 445, 1982.
6. **Ciapetta, F. G., Dobres, R. M., and Baker, R. W.,** in *Catalysis,* Vol. 6, Emmett, P. H., Ed., Reinhold, New York, 1958.
7. **Schuit, G. C. A. and Gates, B. C.,** *Chem. Tech,* 556, September 1983.
8. **Gates, B. C., Katzer, J. R., and Schuit, G. C. A.,** *Chemistry in Catalytic Processes,* McGraw-Hill, New York, 1979.
9. **Kluksdahl, H. E.,** U. S. Patent 3,419,737, 1968.
10. **Sinfelt, J. H.,** *Bimetallic Catalysts,* John Wiley & Sons, New York, 1983, chap. 5; and references cited therein.
11. **Mc Callister, T. P. and O'Neal, K. R.,** German Patent 2,104,429, 1971.
12. **Dautzenberg, F. M.,** German Patent 2,121,265, 1971.
13. **Dautzenberg, F. M. and Kouwenkoven, H. W.,** German Patent 2,153,891, 1972.
14. **Srivastava, R. D. and Farber, M.,** *Chem. Rev.,* 78, 627, 1978.
15. **Svajgl, O.,** *Int. Chem. Eng.,* 12(1), 55, 972.
16. **Anon, A.,** *Oil Gas J.,* 64, 87, 1966.
17. **Prins, R.,** in *Chemistry and Chemical Engineering of Catalytic Processes,* Prins, R. and Schuit, G. C. A., Eds., Sijthoff and Noordhoff, Alphen aan den Rijn, Netherlands, 1980, 389.
18. **Ciapetta, F. G. and Wallace, D. N.,** *Catal. Rev.,* 5, 67, 1971; and references cited therein.
19. **Anderson, J. R.,** *Adv. Catal.,* 23, 1, 1973.
20. **Clarke, J. K. A. and Rooney, J. K.,** *Adv. Catal.,* 25, 125, 1976.
21. **Gault, F. G.,** *Adv. Catal.,* 30, 1, 1981.
22. **Ponec, V.,** *Adv. Catal.,* 32, 144, 1983.
23. **Weisz, P. B.,** *Adv. Catal.,* 13, 137, 1962; and references cited therein.
24. **Mills, G. A., Heinmann, H., Milliken, T. H., and Obland, A. G.,** *Ind. Eng. Chem.,* 45, 134, 1953.
25. **Hindin, S. G., Weller, S. W., and Mills, G. A.,** *J. Phys. Chem.,* 62, 244, 1958.
26. **Sinfelt, J. H., Hurwitz, H., and Rohrer, J. C.,** *J. Phys. Chem.,* 62, 244, 1958.
27. **Anderson, J. R. and Baker, B. G.,** *Proc. R. Soc. London Ser. A,* 271, 402, 1963.
28. **Carter, J. L., Cusumano, J. A., and Sinfelt, J. H.,** *J. Catal.,* 20, 223, 1971.
29. **Boudart, M. and Ptak, L. D.,** *J. Catal.,* 20, 223, 1971.
30. **Matsumoto, H., Saito, Y., and Yoneda, Y.,** *J. Catal.,* 19, 101, 1970.

31. **Anderson, J. R. and Avery, N. R.,** *J. Catal.*, 19, 41, 1970.
32. **Dautzenberg, F. M. and Platteeuw, J. C.,** *J. Catal.*, 19, 41, 1970.
33. **Sinfelt, J. H.,** *Adv. Catal.*, 23, 91, 1973.
34. **Sinfelt, J. H. and Rohrer, J. C.,** *J. Phys. Chem.*, 65, 2272, 1961; 66, 1559, 1962.
35. **Spenadel, L. and Boudart, M.,** *J. Phys. Chem.*, 64, 204, 1960.
36. **Hall, W. K. and Lutinski, F. E.,** *J. Catal.*, 2, 518, 1963.
37. **Benson, J. E. and Boudart, M.,** *J. Catal.*, 4, 704, 1965.
38. **Adams, C. R., Banesi, H. A., Curtis, R. M., and Meisenheimer, J. R.,** *J. Catal.*, 1, 336, 1962.
39. **Wilson, G. R. and Hall, W. K.,** *J. Catal.*, 17, 190, 1970.
40. **Via, G. H., Sinfelt, J. H., and Lytle, F. W.,** *J. Chem. Phys.*, 71, 690, 1979.
41. **Monogue, W. H. and Katzer, J. R.,** *J. Catal.*, 32, 166, 1974.
42. **Barron, Y., Maire, G., Muller, J. M., and Gault, F. G.,** *J. Catal.*, 5, 428, 1966.
43. **Maire, G., Plouidy, G., Prudhomme, J. C., and Gault, F. G.,** *J. Catal.*, 4, 556, 1965.
44. **Dautzenberg, F. M. and Platteeuw, J. C.,** *J. Catal.*, 24, 264, 1972.
45. **Maire, G., Corolleur, C., Juttard, D., and Gault, F. G.,** *J. Catal.*, 21, 250, 1971.
46. **Lankhorst, P. P., Dejongste, H. C., and Ponec, V.,** in *Catalyst Deactivation*, Delmon, B. and Froment, G. C., Eds., Elsevier, Amsterdam, 1980.
47. **Santacessaria, E., Gelosa, D., Carra, S., and Adami, I.,** *Ind. Eng. Chem. Prod. Res. Dev.*, 17, 68, 1968.
48. **Anderson, J. R. and Shimoyama, Y.,** Proc. 5th Int. Congr. Catal., Miami, 1972, preprint.
49. **Corolleur, C., Tamanova, D., and Gault, F. G.,** *J. Catal.*, 24, 401, 1972.
50. **Anderson, J. R., Macdonald, R. J., and Shimoyama, Y.,** *J. Catal.*, 20, 147, 1971.
51. **Davis, S. M., Zarera, F., and Somorjai, G. A.,** *J. Catal.*, (a) 77, 439, 1982; (b) 85, 206, 1984.
52. **Davis, S. M. and Somorjai, G. A.,** *J. Phys. Chem.*, 87, 1545, 1983.
53. **Sinfelt, J. H., Via, G. H., and Lytle, F. W.,** *J. Chem. Phys.*, 76, 2779, 1982.
54. **Sinflet, J. H., Via, G. H., and Lytle, F. W.,** *Catal. Rev. Sci. Eng.*, 26, 81, 1984; and references cited therein.
55. **Garten, R. H. and Sinfelt, J. H.,** *J. Catal.*, 62, 127, 1980.
56. **Sinfelt, J. H.,** *Sci. Am.*, 253, 90, 1985.
57. **Sinfelt, J. H., Via, G. H., Meitzner, G., and Lytle, F. W.,** in *Catalyst Characterization Science*, Deviney, M. L. and Gland, J. L., Eds., ACS Symposium Series, No. 288, 254, 1985.
58. **Lytle, F. W., Greegor, R. B., Marques, E. C., Biebesheimer, V. A., Sandstrom, D. R., Horsley, J. A., Via, G. H., and Sinfelt, J. H.,** in *Catalyst Characterization Science*, Deviney, M. L. and Gland, J. L., Eds., ACS Symposium Series, No, 288, 281, 1985.
59. **Sinfelt, J. H.,** *J. Phys. Chem.*, 90, 4711, 1986.
60. **Sinfelt, J. H.,** *Ind. Eng. Chem. Fundam.*, 25, 2, 1986.
61. **Johnson, M. F. and Le Roy, V. M.,** *J. Catal.*, 35, 434, 1974.
62. **Johnson, M. F.,** *J. Catal.*, 39, 487, 1975.
63. **Webb, A. N.,** *J. Catal.*, 34, 485, 1975.
64. **Menon, P. G., Sieders, J., Streefkerk, F. J., and Van Keulen, G. J. M.,** *J. Catal.*, 29, 188, 1973.
65. **Freel, J.,** *Am. Chem. Soc. Div. Pet. Chem. Prepr.*, 18, 10, 1973, preprint.
66. **Bolivar, C., Charcosset, H., Frety, R., Primet, M., Tournayan, L., Betizeau, C., Leclercq, G., and Maurel, R.,** *J. Catal.*, 39, 249, 1975.
67. **Betizeau, C., Bolivar, C., Charcosset, H., Frety, R., Leclercq, G., Maurel, R. and Tournayan, L.,** in *Preparation of Catalysts*, Delmon, B., Jacobs, P. A., and Poncelet, C., Eds., Elsevier, Amsterdam, 1976, 525.
68. **Yao, H. C. and Shelef, M.,** *J. Catal.*, 44, 392, 1976.
69. **Barbier, J., Charcosset, H., Periera, G., and Riviere, J.,** *Appl. Catal.*, 1, 71, 1981.
70. **Apestguia, C. R. and Barbier, J.,** *J. Catal.*, 78, 352, 1982.
71. **Wagstaff, N. and Prins, R.,** *J. Catal.*, 59, 434, 1979.
72. **Mc Nicol, B. D.,** *J. Catal.*, 46, 438, 1977.
73. **Betizeau, C., Leclercq, G., Maurel, R., Bolivar, G., Charcosset, H., Frety, R., and Tournayan, L.,** *J. Catal.*, 45, 179, 1976.
74. **Charcosset, H., Frety, R., Leclercq, G., Mendes, E., Primet, M., and Tournayan, L.,** *J. Catal.*, 56, 468, 1979.
75. **Shiflett, W. K. and Dumesic, J. A.,** *J. Catal.*, 77, 57, 1982.
76. **Cimino, A., De Angelis, B. A., Gazzoli, D., and Valigi, M.,** *Z. Anorg. Allg. Chem.*, 460, 86, 1980.
77. **Biloen, P., Helle, J. N., Verbeek, H., Dautzenberg, F. M., and Sachtler, W. M. H.,** *J. Catal.*, 63, 112, 1980.
78. **Isaac, B. H. and Petersen, E. E.,** *J. Catal.*, 77, 43, 1982.
79. **Peri, J. B.,** *J. Catal.*, 52, 144, 1978.

80. **Bertolacini, R. J. and Pellet, R.,** in *Catalyst Deactivation,* Delmon, B. and Froment, G. F., Eds., Elsevier, Amsterdam, 1980, 73.
81. **Goldstein, M. S.,** in *Experimental Methods in Catalyst Research,* Anderson, R. B., Ed., Academic Press, New York, 1968, 361.
82. **Buss, W. C. and Hughes, T. R.,** U.S. Patent 4,435,283, 1984; Belgian Patent 895,778, 1983.
83. **Freed, R. L. and Kelley, M. J.,** in Proc. 38th Annu. Meet. Elect. Micros. Soc. Am., 1980, 200.
84. **Short, D. R., Khalid, S. M., Katzer, J. R., and Kelley, M. J.,** *J. Catal.,* 72, 288, 1981.
85. **Onuferko, J., Short, D. R., and Kelley, M. J.,** *Appl. Surf. Sci.,* 19, 227, 1984.
86. **McMaster, W. H., Del Grande, N. K., Mallet, J. H., and Hubbell, J.,** Compilation of X-ray Cross Section, Lawrence Radiation Laboratory, Livermore, California, 1969.
87. **Sinfelt, J. H.,** U.S. Patent 3,953,368, 1976.
88. **Wagstaff, N. and Prins, R.,** *J. Catal.,* 59, 446, 1979.
89. **Sinfelt, J. H. and Via, G. H.,** *J. Catal.,* 56, 1, 1979.
90. **Davis, B. H.,** *J. Catal.,* 46, 348, 1977.
91. **Leclercq, G., Jaunay, D., and Maurel, R.,** *J. Chem. Res. Miniprint,* 716, 1979.
92. **Dautzenberg, F. M., Hella, J. N., Biloen, P., and Sachtler, W. M. H.,** *J. Catal.,* 63, 119, 1980.
93. **Burch, R. and Garcia, L. C.,** *J. Catal.,* 71, 360, 1981.
94. **Volter, J., Leitz, G., Uhlemann, M., and Hermann, M.,** *J. Catal.,* 68, 42, 1981.
95. **Coq, B. and Figueras, F.,** *J. Catal.,* 85, 197, 1984.
96. **Sexton, B. A., Hughes, A. E., and Forger, K.,** *J. Catal.,* 88, 466, 1984.
97. **Hughes, T. R., Buss, W. C., Tamm, P. W., and Jacobson, R. L.,** in *Proc. 7th Int. Zeolite Conf.,* Murakami, Y., Lijima, A., and Ward, J. W., Eds., Kodansha, Tokyo, 1986, 725; and references cited therein.
98. **Bouwman, R., Toneman, L. H., and Holecher, J. E.,** *Surf. Sci.,* 35, 8, 1973.
99. **Bouwman, R. and Biloen, P.,** *Surf. Sci.,* 41, 348, 1974.
100. **Berndt, H., Mehner, H., Volter, J., and Meisel, W.,** *Z. Anorg. Allg. Chem.,* 429, 27, 1977.
101. **Burch, R.,** *J. Catal.,* 71, 348, 1981.
102. **Lieske, H. and Volter, J.,** *J. Catal.,* 90, 96, 1984.
103. **Adkins, A. R. and Davis, B. H.,** *J. Catal.,* 89, 371, 1984.
104. **Muller, A., Englard, P. A., and Weisang, J. E.,** *J. Catal.,* 56, 65, 1979.
105. **Baonetti, G. T., Miguel, S. R., Scelza, O. A., Fritzler, M. A., and Castro, A. A.,** *Appl. Catal.,* 19, 77, 1985.
106. **Bacaud, R., Bussier, P., and Figueras, F.,** *J. Catal.,* 69, 399, 1981.
107. **Liu, N., Yang, X., and Pang, L.,** *Chin. J. Catal.,* 6, 13, 1985.
108. **Sachtler, W. M. H.,** *J. Mol. Catal.,* 25, 1, 1984.
109. **Biloen, P., Helle, J. N., Verbeek, H., Dautzenberg, F. M., and Sachtler, W. M. H.,** *J. Catal.,* 63, 112, 1980.
110. **Srivastava, R. D., Prasad, N. S., and Pal, A. K.,** *Chem. Eng. Sci.,* 41, 719, 1986.
111. **Srivastava, R. D., Sengupta, S. K., Johri, U. C., Singru, R. M., and Kunzru, D.,** to be published.
112. **Carter, J. L., McVicker, G. B., Weissman, W., Kmak, W. S., and Sinfelt, J. H.,** *Appl. Catal.,* 3, 327, 1982.
113. **McVicker, G. B. and Ziemiak, J. J.,** *Appl. Catal.,* 14, 229, 1985.
114. **Haensel, V. and Slerba, M. J.,** *Ind. Eng. Chem. Prod. Res. Dev.,* 15, 3, 1976.
115. **Menon, P. G. and Prasad, J.,** *Proc. 6th Int. Congr. Catal.,* Vol 2, Bond, G. C., Wells, P. B., and Tompkins, F. C., Eds., Chemical Society, London, 1977, 1061.
116. **Parera, J. M., Beltramini, J. N., Querini, C. A., Matinelli, E. E., Churin, E. J., Aloe, P. E., and Figoli, N. S.,** *J. Catal.,* 99, 39, 1986; and references cited therein.
117. **Boliver, C., Charcosset, H., Frety, R., Primet, M., Tournayan, L., Betizeau, C., Leclercq, G., and Maurel, R.,** *J. Catal.,* 45, 163, 1976.
118. **Jossens, L. W. and Petersen, E. E.,** *J. Catal.,* 76, 265, 1982.
119. **Jothimurugesan, K., Bhatia, S., and Srivastava, R. D.,** *Ind. Eng. Chem. Fundam.,* 24, 433, 1985.
120. **Pal, A. K., Bhowmick, M., and Srivastava, R. D.,** *Ind. Eng. Chem. Process Des. Dev.,* 25, 236, 1986.
121. **Van Trimpont, P. A., Marin, G. B., and Froment, G. F.,** *Ind. Eng. Chem. Fundam.,* 25, 544, 1986.
122. **Kelley, M. J. and Dadyburjor, D. B.,** 3rd Int. Conf. on Catalyst, Sintering and Deactivation, Berkeley, 1985, preprint.
123. **Ludlum, K. H. and Eischens, R. P.,** *Am. Chem. Soc. Div. Pet. Chem. Prepr.,* 21, 2, 1976, preprint.
124. **Espinat, D., Dexpert, H., Freund, E., Martino, G., Couzi, M., Lespade, D., and Cruege, F.,** *Appl. Catal.,* 16, 343, 1985.
125. **Parera, J. M., Jablonski, E. L., Cabrol, R. A., Figoli, N. S., Musso, J. C., and Verderone, R. J.,** *Appl. Catal.,* 12, 125, 1984.
126. **Parera, J. M., Figoli, N. S., Beltramini, J. N., Churin, E. J., and Cabrol, R. A.,** *Proc. 8th Int. Congr. Catal.,* Vol. 2, Verlag Chemie, Berlin, 1984, 593.

127. **Beltramini, J. N., Churin, E. J., Traffano, E. M., and Parera, J. M.,** *Appl. Catal.*, 19, 203, 1985.
128. **Parera, J. M., Beltramini, J. N., Querini, C. A., and Figoli, N. S.,** Am. Chem. Soc. Div. Pet. Chem., p. 532, 1985.
129. **Verderone, R. J., Pieck, C. L., Sad, M. R., and Parera, J. M.,** *Appl. Catal.*, 21, 239, 1986.
130. **Parera, J. M., Verderone, R. J., Pieck, C. L., and Traffano, E. M.,** *Appl. Catal.*, 23, 15, 1986.
131. **Beltramini, J. N. and Parera, J. M.,** *Appl. Catal.*, 7, 43, 1983.
132. **Shum, V. K., Butt, J. B., and Sachtler, W. M. H.,** *Appl. Catal.*, 11, 151, 1984.
133. **Chrstoffel, E., Petting, P., and Vierrath, H.,** *J. Catal.*, 40, 349, 1975.
134. **Carter, J. L. and Sinfelt, J. H.,** U.S. Patents 3,939,062, 1976; 3,940,510, 1976.
135. **(a) Yates, D. J. C.,** U.S. Patent 3,981,823, 1976
(b) Kmak, W. S. and Yates, D. J. C., U.S. Patent 3,943,052, 1976.
136. **(a) Fung, S. C., Weissman, W., and Carter, J. L.,** U.S. Patent 4,444,895, 1984.
(b) Fung, S. C., Weissman, W., Carter, J. L., and Kmak, W. S., U.S. Patent 4,444,897, 1984.
137. **Chester, A. W., Krishnamurthy, S., Kresge, C. T., and McHale, W. D.,** European Patent Appl. 0,103,449, 1983.
138. **Sinfelt, J. H.,** U.S. Patent 3,791,961, 1974.
139. **Pacheco, M. and Petersen, E. E.,** *J. Catal.*, 86, 75, 1984; 88, 400, 1984.
140. **Krane, H. G., Groh, A. B., Schulman, B. L., and Sinfelt, J. H.,** Proc. 5th World Pet. Congr., New York, 1959, 39.
141. **McVicker, G. B., Collins, P. J., and Ziemiak, J. J.,** *J. Catal.*, 74, 156, 1982.
142. **Coughlin, R. W., Hasan, A., and Kawakami, K.,** *J. Catal.*, 80, 150, 1984; 88, 163, 1984.
143. **Ramage, M. P., Graziani, K. R., and Krambeck, F. J.,** *Chem. Eng. Sci.*, 35, 41, 1980.
144. **Smith, R. B.,** *Chem. Eng. Prog.*, 55, 76, 1959.
145. **Henningsen, J. and Bundgaard, N. M.,** *Br. Chem. Eng.*, 15, 1433, 1970.
146. **Bommannan, D.,** Thesis, Indian Institute of Technology, Kanpur, 1986.
147. **Van Der Baan, H. S.,** in *Chemistry and Chemical Engineering of Catalytic Processes,* Prins, R. and Schuit, G. C. A., Eds., Sijthoff and Noordhoff, Alphen aan den Rijn, Netherlands, 1980, 381.

INDEX

S

T

U

V

W

X

Z